P9-CAM-935

BIOLOGICAL CONSEQUENCES OF OXIDATIVE STRESS

THE CONTE INSTITUTE SERIES

1. *Biological Consequences of Oxidative Stress: Implications for Cardiovascular Disease and Carcinogenesis*
   Edited by Lawrence Spatz and Arthur D. Bloom

BIOLOGICAL CONSEQUENCES OF OXIDATIVE STRESS

Implications for Cardiovascular Disease and Carcinogenesis

Edited by

Lawrence Spatz, Ph.D.
Arthur D. Bloom, M.D.

The Conte Institute for Environmental Health

New York Oxford
OXFORD UNIVERSITY PRESS
1992

Oxford University Press

Oxford New York Toronto
Delhi Bombay Calcutta Madras Karachi
Kuala Lumpur Singapore Hong Kong Tokyo
Nairobi Dar es Salaam Cape Town
Melbourne Auckland

and associated companies in
Berlin Ibadan

Published by Oxford University Press, Inc.,
200 Madison Avenue, New York, New York 10016

Library of Congress Cataloging-in-Publication Data
Biological consequences of oxidative stress:implications for cardiovascular disease and carcinogenesis/edited by Lawrence Spatz, Arthur D. Bloom.
p. cm.
Includes bibliographical references and index.
ISBN 0-19-507296-0
1. Active oxygen—Pathophysiology. 2. Carcinogenesis.
3. Cardiovascular system—Pathophysiology. 4. Free radicals (chemistry)—Pathophysiology. I. Spatz, Lawrence.
II. Bloom, Arthur D.
[DNLM: 1. Cardiovascular Diseases—etiology. 2. Free Radicals.
3. Neoplasms—etiology. 4. Oxygen—adverse effects.
5. Oxygen—metabolism.
QV 312 B6152] RB170.B56 1992 616.1′071—dc20 DNLM/DLC
for Library of Congress 91-32070

9 8 7 6 5 4 3 2 1

Printed in the United States of America
on acid-free paper

Preface

The Silvio O. Conte Institute for Environmental Health was established in 1987 to provide a scientific forum for the critical evaluation of basic and applied environmental health research. The Institute provides analyses of scientific data and hypotheses on the effects of environmental agents on human health and disseminates these analyses to other scientists, to non-scientists, and, in general, to those concerned with protection of the public health.

The present monograph was produced by a group of Institute Fellows whose purpose was to evaluate the ways in which environmental agents modulate oxidative damage and to assess the role of oxidative damage in the pathogenesis of cancer and cardiovascular disease. This interdisciplinary study group met several times to discuss, and write critically about, the current state of our knowledge in this field. The group considered: the basic mechanisms by which reactive oxygen species may cause cellular and molecular damage; the interaction of genetic and environmental factors in the production of this damage; the role of oxidative damage in atherosclerosis and in carcinogenesis; the role of iron in oxidative stress; and the modulation of carcinogenesis by antioxidants. The individual chapters of the book reflect per force the viewpoints of the individual authors as tempered and refined by the interactive, critical study group process. We anticipate, then, that this book will serve both to introduce and to summarize oxidative damage as a primary pathophysiological process with important clinical implications.

The work of the study group was funded by the National Institute of Environmental Health Sciences (NIEHS) of the National Institutes of Health (contract number N01-ES-75185). We thank NIEHS project officers Doctors David Hoel and James Huff, and the contract officer, Mr. Phillip Jones, for support throughout this effort. We thank too, Mr. Jeffrey House and Ms. Edith Barry, of Oxford University Press for assisting so ably in the production of this inaugural monograph of The Conte Institute Series.

Pittsfield, Massachusetts
February, 1992

Lawrence Spatz, Ph.D.
Arthur D. Bloom, M.D.

Contents

Contributors

Arthur D. Bloom, M.D.
The Conte Institute for Environmental Health
Pittsfield, Massachusetts 01201

Donna L. Carden, M.D.
Department of Physiology
Louisiana State University Medical Center
Shreveport, Louisiana 71130

Guy M. Chisolm III, Ph.D.
Department of Vascular Cell Biology and Atherosclerosis
Research Institute of The Cleveland Clinic Foundation
Cleveland, Ohio, 44195

Nancy H. Colburn, Ph.D.
Laboratory of Viral Carcinogenesis
National Cancer Institute
Frederick Cancer Research and Development Center
Frederick, Maryland, 21702

D. Neil Granger, Ph.D.
Department of Physiology
Louisiana State Medical Center
Shreveport, Louisiana 71130

Kathryn Z. Guyton, B.A.
Department of Environmental Health Sciences
Johns Hopkins University
School of Hygiene and Public Health
Baltimore, Maryland 21205

Thomas W. Kensler, Ph.D.
Department of Environmental Health Sciences
Johns Hopkins University
School of Hygiene and Public Health
Baltimore, Maryland 21205

Ronald J. Korthuis, Ph.D.
Department of Physiology
Louisiana State University Medical Center
School of Medicine in Shreveport
Shreveport, Louisiana 71130

Stuart Linn, Ph.D.
Department of Molecular and Cell Biology
University of California
Berkeley, California 94720

Ronald P. Mason, Ph.D.
Laboratory of Molecular Biophysics
National Institute of Environmental Health Sciences
Research Triangle Park, North Carolina 27709

Kazuo Neriishi, M.D.
Radiation Effects Research Foundation
Hiroshima, Japan

Lawrence Spatz, Ph.D.
The Conte Institute for Environmental Health
Pittsfield, Massachusetts 01201

Richard G. Stevens, Ph.D.
D. Battelle
Pacific Northwest Laboratory
Richland, Washington 99352

BIOLOGICAL CONSEQUENCES OF OXIDATIVE STRESS

1
Introduction

LAWRENCE SPATZ

Molecular oxygen, although indispensable to aerobic life forms as the final electron acceptor for energy-yielding reactions, is simultaneously a universal toxicant. This fact, often obscured by both the frequency with which oxygen is employed in clinical medicine and its clearly essential role in aerobic metabolism, is borne out by experimental evidence. Exposure of bacteria, yeast, algae, or plants to oxygen concentrations greater than they ordinarily encounter leads to reduced viability, mutagenesis, and growth suppression (Kaye 1967; Gregory et al. 1974; Fenn et al. 1957; Pulich 1974; Rabinowitch and Fridovich 1983). In mammals, hyperoxic exposures lead to at least three types of deleterious effects:

1. Retinal effects. Retrolental fibroplasia occurs in premature infants exposed to oxygen before 44 weeks of gestational age (Betts et al. 1977). It is characterized by slowly developing retinal damage that appears to be a consequence of oxygen-induced spasm and subsequent degeneration of immature retinal arterioles. The damage may regress entirely or may progress to retinal detachment and blindness. It is induced by arterial Po_2 levels in excess of 60 to 70 mm Hg (Ashton 1979).
2. Pulmonary effects. Pulmonary manifestations of oxygen toxicity are those most commonly encountered, because the lungs are the organs most immediately exposed to O_2. Symptoms include irritation of the respiratory tract characterized by a progressive decrease in vital capacity and by coughing, nasal stuffiness, sore throat, substernal distress, and tracheobronchitis; with prolonged exposure, pulmonary congestion and atelectasis also occur (Miller and Winter 1981). Endothelial and type 1 alveolar cells are most susceptible to damage, and the alveolar capillary membrane may lose its integrity, resulting in increased interstitial fluid and protein. In addition, after several hours of hyperoxic exposure, there may be inhibition of the muco-

ciliary transport mechanism and a subsequent impairment of the pulmonary defense mechanisms (Barnes et al. 1983; Davis et al. 1983; Deneke and Fanburg 1982). Increased production of partially reduced forms of oxygen, such as superoxide anion and hydrogen peroxide, has been observed in lung tissue and isolated lung subcellular organelles during hyperoxia.

3. Nervous system effects. Inhalation of pure oxygen at greater than 2 atm produces a characteristic central nervous system (CNS) syndrome that includes mood changes, nausea, vertigo, muscular twitching, convulsions, and loss of consciousness. These CNS changes appear to be reversible with a decrease in Po_2 below 2 atm (Smith et al. 1985). It is important to remember that hemoglobin is nearly saturated with oxygen by air at normal atmospheric pressure and that the major effect of hyperoxic exposures is to increase markedly the amount of oxygen physically dissolved in blood. Under several atmospheres of pure oxygen, the amount of oxygen physically dissolved in the blood is almost equivalent to that carried by hemoglobin (DiGuiseppi and Fridovich 1983). A substantial increase in the amount of oxygen delivered to the tissues can occur, and damage to many organs, such as the heart, kidneys, livers, testes, and bone marrow, can result (Balentine 1978).

Oxygen sensitivity can be modulated by age, diet, and physiologic state. Young mice are less susceptible than adult mice (Bonikos et al. 1976). Adrenalectomy, hypophysectomy, and thyroidectomy decrease susceptibility, whereas administration of adrenocorticotropin, epinephrine, and thyroxine increase it (Taylor 1954). Increased intake of polyunsaturated fatty acids increases susceptibility in rats, as does vitamin E deficiency (Tierney et al. 1977; Kehrer and Autor 1978). Tolerance to hyperoxia can be induced by brief pre-exposure to 85% oxygen. The pre-exposure results in the induction of increased levels of several antioxidant enzymes, such as superoxide dismutase and catalase, which can protect tissues from oxidative damage and thus cause tolerance (Fridovich and Freeman 1986). Indeed, intravenous injection of these enzymes encased in liposomes also produces tolerance (Turrens et al. 1984). Induction of these enzymes is part of an adaptive metabolic response to increased production of partially reduced forms of oxygen in tissues; because the concerted action of these enzymes on oxygen metabolites regenerates molecular oxygen and water, their net effect is to reduce tissue levels of oxygen metabolites. These observations, and many others, have led to the hypothesis that oxygen toxicity is a complex phenomenon mediated not by oxygen itself, but by certain reactive intermediates generated in the process of its reduction and that cells and organisms have developed many defense mechanisms that protect them from damage (DiGuiseppi and Fridovich 1983). There is also considerable evidence that the defense against oxygen toxicity is never completely successful even at atmospheric concentrations of O_2 (21%) and that many disease processes such as those that underlie cancer and heart disease, as well as the aging process itself, are, at least in part, manifestations of the slow accumulation through time of oxidative damage to crucial cellular components.

This book is an in-depth examination of the evidence supporting this concept and a detailed examination of the ways in which oxidative damage contributes to heart disease and carcinogenesis. It also examines the ways in which environmental exposures can modulate oxidative damage.

THE CHEMISTRY OF MOLECULAR OXYGEN

The complete reduction of molecular oxygen to water requires the addition of four electrons ($O_2 + 4H^+ + 4e^- \longrightarrow 2H_2O$), but because of its molecular structure, oxygen is usually constrained to receive these electrons one at a time. Technically speaking, oxygen has two unpaired electrons in its molecular orbitals, and these two electrons have the same spin quantum number (parallel spin). If O_2 oxidizes another atom or molecule by accepting a pair of electrons, both of these electrons must be of parallel spin. Most biomolecules however, are covalently bonded nonradicals, and the electron pairs in their covalent bonds have opposite spins (antiparallel spin) and occupy the same molecular orbital. The donation of a pair of antiparallel spin electrons from the molecular orbital of a biomolecule to molecular oxygen is thus spin restricted, and, consequently, oxygen is more likely to undergo single-electron reductions. This spin restriction is of considerable importance to biologic systems because it prevents the large pools of carbon- and hydrogen-containing biomolecules from spontaneously combusting in the earth's oxygen-rich environment.

As a consequence of oxygen's tendency to undergo reduction in single-electron steps, several intermediates arise in the process that are of considerable interest and reactivity. It is these intermediates, usually designated as reactive oxygen species (ROS) or activated oxygen species (AOS), that are believed to be the cause of oxygen's toxicity and of "oxidative stress." The first of these is the superoxide radical, $\dot{O}_2^-$, arising by the addition of one electron and produced in virtually all aerobic cells. The superoxide radical in hydrophobic environments (as in the interior of biologic membranes) is a powerful base (proton acceptor), nucleophile, and reducing agent (Frimer 1982; Valentine et al. 1984). It can also act as an oxidizing agent, but only with compounds that can serve as proton donors (such as ascorbate or vitamin E). O_2 generated within the hydrophobic interior of membranes can cause considerable damage by destroying phospholipids. Because oxygen is more soluble in hydrophobic environments than in water and because many of the $\dot{O}_2^-$-generating systems of cells reside in membranes, it is likely that a substantial amount of the biologic generation of superoxide occurs at these sites and brings with it the potential for membrane damage (Halliwell and Gutteridge 1984). In an aqueous milieu, superoxide undergoes two major reactions: It acts as a single-electron reducing agent, and it reacts with itself (disproportionates or dismutates) to produce O_2 and H_2O_2, hydrogen peroxide: $2\dot{O}_2^- + 2H^+ \longrightarrow H_2O_2 + O_2$. This latter reaction is more rapid at acidic pH and occurs spontaneously by way of biomolecular collision between superoxide molecules. Despite the occurence of spontaneous dismutation of superoxide, in many biologic systems, enzymes, or superoxide dismutases, are present that can substantially accelerate the reaction rate and reduce the half-life of superoxide radicals in tissues.

The dismutation of superoxide generates another ROS, hydrogen peroxide. Hydrogen peroxide does not have any unpaired electrons and is not a radical, but there are two ways in which it can create problems. First, it is water soluble, it can cross biologic membranes (which superoxide cannot except through anion channels), and, because of its limited reactivity, it can be widely dispersed. Second, in the presence of ferrous iron (Fe^{++}) or other transition metals, it can accept a single electron to produce the highly reactive hydroxyl radical, $\dot{O}H$: $Fe^{++} + H_2O_2 \longrightarrow Fe^{+++} + \dot{O}H + OH^-$. The ferric ion ($Fe^{+++}$) can be re-reduced to ferrous (Fe^{++}) by either H_2O_2 or superoxide: $Fe^{+++} + H_2O_2 \longrightarrow Fe^{++} + \dot{O}_2^- + 2H^+$; $Fe^{+++} + \dot{O}_2^- \longrightarrow Fe^{++} + O_2$. Other one-electron transfers are also possible:

$$\dot{O}H + H_2O_2 \longrightarrow H_2O + H^+ + \dot{O}_2^-$$

$$\dot{O}H + Fe^{++} \longrightarrow Fe^{+++} + OH^-$$

The generation of hydroxyl radical by the simple mixture of hydrogen peroxide and Fe^{++} was first observed by Fenton in 1893. These iron-catalyzed reactions in which H_2O_2 is converted to water and molecular oxygen with generation of hydroxyl radicals as intermediates is now frequently referred to as Fenton chemistry. It is important to remember that even traces of iron can be effective generators of hydroxyl radicals because the iron acts catalytically as it cycles between ferrous and ferric forms by reacting with H_2O_2 and $\dot{O}_2^-$ (see preceding explanation). Although the rate constant for iron-catalyzed hydroxyl radical production from H_2O_2 is not high, it has been estimated from the concentration of the reactants in liver cells, that about 50 hydroxyl radicals are produced every second in each liver cell by way of Fenton reactions (Halliwell and Gutteridge 1990). There is yet another source of hydroxyl radicals: One of the major effects of ionizing radiation in biologic systems is the breakdown of water molecules to a variety of reactive species, among which is the hydroxyl radical.

The hydroxyl radical reacts with almost all types of biomolecules—sugars, amino acids, DNA bases, phospholipids, organic acids—diffusing only 5 to 10 molecular diameters from its site of formation before reacting and causing damage by either abstracting a hydrogen atom, adding on to an existing molecule, or transferring an electron (Halliwell and Gutteridge

1984). The hydroxyl radical is of such reactivity that, whenever it is formed in a biologic system, it reacts immediately with whatever biomolecule is in its vicinity, damaging it and producing secondary radicals of varying reactivities. Thus, if a superoxide radical is generated and iron or another transition metal ion is available, there is a significant potential for damage through the hydroxyl radical.

The reactivity of molecular oxygen can also be increased by moving one of the unpaired electrons in such a manner as to alleviate the spin restriction. This event requires the input of energy—usually provided by light acting on pigments such as chlorophyll or porphyrins—and most commonly generates singlet states of oxygen such as $^1\Delta gO_2$. Singlet O_2 formation occurs in pigment-containing illuminated systems, such as the lens of the eye and chloroplasts of plant leaves (Zigler and Goosey 1981; Halliwell 1984); it may also be of significance as a cause of the dermal symptoms (eruptions, scarring, and thickening) seen in some types of porphyria in which particular protoporphyrin IX precursors accumulate in the skin (Halliwell and Gutteridge 1990). Some drugs (for example, psoralens, tetracycline, benoxaprofen) induce photosensitization of the skin and may do so by increasing singlet O_2 production (Motten and Chignell 1983). It has been suggested that singlet oxygen is generated during the dismutation of superoxide radicals and the respiratory burst of neutrophils, but the evidence for this has been questioned. The exact role of singlet oxygen in the biologic toxicity of oxygen is at present unclear, but it may be involved in lipid peroxidation.

Because molecular oxygen has two unpaired electrons of parallel spin, its chemistry is dominated by reactive intermediates whose production is enhanced by transition metal ions, particularly iron. In aerobic biologic systems, iron is generally present in substantial quantity as a necessary constituent of the oxygen-transport protein, hemoglobin, and because of its ability as a constituent of various enzymatic proteins (for example, cytochrome a_3) to overcome oxygen's spin restriction. Superoxide radicals are constantly generated from several sources: the leakage of electrons to oxygen from the various cellular electron transport enzymes in mitochondria and endoplasmic reticulum; the respiratory burst of neutrophils, monocytes, macrophages, and eosinophils; and the metabolism of xenobiotics. It has been estimated that up to 5% of the total O_2 reduced is converted to superoxide (Boveris and Cadenas 1982). Clearly then, without some means of protection—some kind of antioxidant defense—aerobic biologic systems are at constant risk of damage from reactive oxygen species. Even when defenses are available, some damage may be inevitable because of the extreme reactivity of hydroxyl radicals; that is, there may be site-specific generation of hydroxyl radicals because of the presence of bound metal ions that can cause damage that is not preventable by antioxidants acting in either aqueous or hydrophobic bulk phases. Indeed, it has been suggested that the sites most susceptible to hydroxyl radical attack may be those at which catalytic metal ions are bound.

THE ANTIOXIDANT DEFENSE

The requirements for an adequate antioxidant defense should be clear from the foregoing discussion: limitation on the amounts of iron and other transition metal ions present to what is needed and their sequestration, as far as possible, in nonreactive forms; means of minimizing the concentrations of superoxide and hydrogen peroxide; and provision for effective scavengers in both aqueous and hydrophobic media that preferentially neutralize radicals that contribute to the generation of ROS. An additional possible response is the evolution of repair systems that specifically correct the damage inflicted on critical cellular structures (such as DNA) by ROS. There is evidence for the existence of all of these in aerobic living systems.

IRON METABOLISM

The metabolism of iron in humans is complex but its general features are as follows: There is a mucosal block to the absorption of iron so that only a small percentage of dietary iron is taken up by the duodenal and jejunal cells at which absorption occurs. Heme from animal sources is absorbed in this way (Morris 1983), but other forms of iron require solubilization and reduction to the ferrous state, which is facilitated by gastric HCl and dietary vitamin C (a reducing agent). Not all of the iron taken up by the mucosal cells is transferred to the circulation; some of it is retained by the cells and is lost when they are sloughed. The average adult man in iron balance absorbs (from the diet) and loses (in sweat, feces, and urine) about 1 mg of iron per day, which constitutes only about 0.02% of the total body iron content. There are extremely efficient means for preserving iron in the body and no known specific physiologic mechanisms for its excretion (Halliwell and Gutteridge 1990).

Once absorbed, most iron is protein-bound. About two thirds of the total iron is present in hemoglobin, and an additional 10% is found in a similar muscle protein, myoglobin. Transferrin is the plasma transport glycoprotein that, at neutral pH, tightly binds two moles of ferric iron per mole of protein. It is ordinarily only about 30% saturated, so that in plasma virtually all of the iron is bound and unreactive. Lactoferrin, a similar protein secreted by activated phagocytes, is present in milk and other body fluids. It is presumed to play a role in preventing oxidative damage at sites of phagocyte activation (for example, infections) (Halliwell and Gutteridge 1990). The intracellular iron-chelating protein is ferritin; it consists of a protein shell capable of binding large amounts of ferric iron (4500 mol iron/mol protein) in its interior as a hydrated ferric oxide-phosphate complex, an unreactive form. In lysosomes, ferritin is converted to an insoluble product known as hemosiderin (Weir et al. 1984).

There exists, in addition to these protein-bound forms of iron a small pool of low-molecular-weight iron chelates whose exact nature is unclear. Some may be attached to DNA; some, to polar head groups of lipids; some, to organic acids (for example, citrate); and some, to the phosphate groups of nucleotides (ATP, ADP, GTP and so forth). These iron chelates are capable of reducing peroxide to hydroxyl radical and can act as catalysts of oxidative damage.

The availability of catalytic iron is thus minimized by the combined actions of restricted absorption and protein sequestration. Under normal conditions, little iron capable of catalyzing oxidative damage is detectable in human body fluids, even by the extremely sensitive bleomycin assay (Gutteridge et al. 1981) specifically designed for its detection. These protective measures can, however, be abrogated when the generation of ROS is accelerated, which occurs when cells are injured or during metabolism of some xenobiotics. Under these conditions, sufficient superoxide ion can be produced to mobilize the iron sequestered in ferritin; at high enough concentrations, hydrogen peroxide can react with hemoglobin and myoglobin to generate reactive species not related to the hydroxyl radical, but nonetheless capable of degrading deoxyribose and stimulating lipid peroxidation (Puppo and Halliwell 1988; Itabe et al. 1988). At higher H_2O_2 concentrations, heme degradation, with release of free iron, occurs (Gutteridge 1986). The potential for liberation of catalytic iron is ever present. It may be of particular significance in certain organs, such as the brain, where the extracellular fluid (cerebrospinal fluid) lacks iron-binding capacity (Bleijenberg et al. 1971). In animal models, the release of iron and subsequent lipid peroxidation induced by brain damage has been noted, as has the diminution of post-traumatic degeneration in the brain and spinal cord by iron-chelating agents (Hall 1988).

Hereditary hemochromatosis (HH) is an iron-overload disease, inherited as an autosomal recessive trait, in which the mucosal block to iron absorption is defective. Heterozygote carriers have a partial defect. There is excessive deposition of iron, as hemosiderin, particularly in the liver, and increased risk of cirrhosis, hepatic and other cancers, and arthritis. Free, catalytically active iron, mostly in the low-molecular-weight citrate form, is detectable by the bleomycin assay in the plasma of patients with HH, and the concentrations decline as the iron overload is brought under control by phlebotomy. It has recently been suggested that the disease is caused by a defect in the ability to attach iron to transferrin in the gut and the subsequent overabsorption of low-molecular-weight forms of iron (Aruoma et al. 1988). The higher risk of cancer in patients with HH may be a consequence of iron-catalyzed oxidative damage to DNA. A more thorough discussion of the association between iron levels and cancer risk is provided in Chapter 7.

COPPER METABOLISM

The other transition metal of interest as a possible catalyst of ROS reactions is copper. Although copper is more reactive than iron, the amount of copper present in humans is about 50-fold less than the amount of iron. Copper is also extensively protein-bound, both to albumin and to the major copper-containing protein of plasma, ceruloplasmin. Copper also binds to amino acids, such as histidine, and to small peptides. Whether copper is available to catalyze the generation of hydroxyl radicals is presently unclear (Halliwell and Gutteridge 1990). The possibility of in situ generation of hydroxyl radicals with damage to the binding molecule has been raised. If the binding molecules are proteins (such as albumin), they may serve a sacrificial protective function by preventing damage to more important targets. On the other hand, copper bound to DNA or to lipids could cause genetic or membrane damage.

ENZYMATIC ANTIOXIDANTS

Regulation of the concentrations of superoxide radicals and hydrogen peroxide in aerobic cells to acceptable levels is accomplished by several enzymatic mechanisms, some of which have been previously alluded to. Measurements of tissue steady-state levels of ROS are not easily obtained, but the H_2O_2 content of the liver has been estimated to be in the range of 10^{-7} to 10^{-9} M whereas the $\dot{O}_2^-$ content is thought to be about 10^{-11} M. The amounts in other tissues may be higher or lower, depending on their metabolic activity and on the effectiveness of their disposal mechanisms, but in most normal cells the ratio of H_2O_2 to $\dot{O}_2^-$ is thought to be about 1000:1 (Cadenas 1989). In addition to its production via the dismutation of $\dot{O}_2^-$, peroxide is also formed directly by several peroxisomal enzymes such as uricase, D-amino acid oxidase, and glycolate oxidase. The enzymes responsible for controlling tissue levels of H_2O_2 and $\dot{O}_2^-$ are superoxide dismutase (SOD), catalase, and glutathione peroxidase.

The SODs are a group of enzymes that catalyze the conversion of $\dot{O}_2^-$ to peroxide and oxygen: $2\dot{O}_2^- + 2H^+ \xrightarrow{SOD} O_2 + H_2O_2$.

As previously noted, this reaction can occur spontaneously through bimolecular collision of two $\dot{O}_2^-$ molecules. For it to occur at an appreciable rate, the bimolecular reaction requires unacceptably high concentrations of superoxide, concentrations that could be detrimental. The SOD enzymes react with individual $\dot{O}_2^-$ molecules at an extremely rapid rate and speed up the dismutation reaction remarkably, thus lowering tissue concentrations of $\dot{O}_2^-$ (Fridovich 1974). There are two types of superoxide dismutases in human tissues: a cytoplasmic copper-zinc dimeric enzyme (molecular weight, 33 kD) and a manganese-containing tetrameric enzyme (molecular weight, 85 kD), present in the mitochondrial matrix. Increased SOD levels have been reported in response to Po_2 increases in several tissues, such as endothelium, rat lung, and alveolar macrophages (Fridovich 1975; 1978). There is as well abundant evidence from a variety of plant and bacterial species that SOD levels are generally reflective of the levels of superoxide radical to which the organism is exposed, increasing and decreasing in direct relationship to exposure (DiGuiseppi and Fridovich 1983).

Catalase is a large (molecular weight, 240 kD) heme-containing tetrameric enzyme found mainly in peroxisomes and mitochondria and probably present in all tissues (Aebi and Wyss 1978). It can catalyze the conversion of H_2O_2 to O_2 and water: $2H_2O_2 \xrightarrow{Cat} O_2 + 2H_2O$. The reaction sequence is initiated by the binding of one molecule of H_2O_2 to the enzyme to form compound I:

Catalase + $H_2O_2 \longrightarrow$ Catalase H_2O_2. Compound I (catalase H_2O_2) then reacts with a suitable electron donor to regenerate the free enzyme: Catalase H_2O_2 + RHOH $\longrightarrow$ Catalase + RO + $2H_2O$.

If H_2O_2 concentrations are high enough ($> 10^{-6}$ M), then H_2O_2 can act as the electron donor as well, yielding the preceding catalatic equation. At low concentrations of H_2O_2, however, alternate electron donors, such as methanol or formic acid, are used, and the enzyme acts as a peroxidase. Because, as noted previously, tissue concentrations of H_2O_2 are thought to be in the range of 10^{-7} to 10^{-9} M, the main physiologic role of catalase is as a peroxidase that uses H_2O_2 as an electron acceptor. It serves a back-up role in antioxidant defense when the levels of H_2O_2 imposed on the cell become excessive. This description of the role of catalase is consistent with the general lack of symptomatic effects seen in the rare genetic condition acatalasemia, which has been described sporadically in Japanese, Korean, Chinese, and Swiss kindreds and is transmitted as an autosomal recessive trait (Aebi and Wyss 1978). Some affected individuals suffer from severe oral ulcers (Takahara's disease), which are caused by the production of excess H_2O_2 by oral bacteria such as hemolytic streptococci and type I pneumococci. In the absence of catalase, the bacterially generated peroxide oxidizes hemoglobin to methemoglobin, which deprives the infected oral areas of oxygen and results in ulceration, necrosis, and tissue decay. Although residual catalase activity can be detected in the tissues of many patients with acatalasemia, some appear to have none and nonetheless appear asymptomatic apart from oral gangrene. Whether these individuals show hypersensitivity to oxidative stress, such as might arise from exposure to radiation or certain chemicals, does not appear to have been investigated.

The major enzymatic system for the control of cellular peroxide levels consists of glutathione (GSH) peroxidases and several ancillary enzymes required for the synthesis and reduction of glutathione. There are at least two forms of glutathione peroxidase, one of which is a selenium-containing tetrameric enzyme (molecular weight, 84 kD), with one selenium atom bound to a cysteine in each subunit (Forstrom et al. 1978). The reaction catalyzed by the enzyme results in the reduction of H_2O_2 to water with generation of oxidized glutathione (GSSG): 2 GSH + $H_2O_2 \longrightarrow$ GSSG + $2H_2O$. Reduced GSH can be regenerated from GSSG by the action of glutathione reductase, a dimeric flavoprotein (molecular weight, 115 kD), which catalyzes the transfer of electrons from NADPH to GSSG (Staal et al. 1969): GSSG + NADPH + $H^+ \longrightarrow$ 2GSH + $NADP^+$. Glutathione can thus cycle between reduced and oxidized forms, reducing H_2O_2 to water at the expense of NADPH. Reduction of $NADP^+$ to NADPH occurs on a continuing basis by the hexose monophosphate shunt in most cells. Glutathione peroxidase catalyzes the reduction of other molecules as well, such as alkyl hydroperoxides (ROOH), to the corresponding alcohols (ROH), and glutathione itself can also react nonenzymatically with H_2O_2 and a variety of free radicals. Glutathione and the enzymes associated with it are critical parts of the antioxidant defense. Further discussion of the reactions of glutathione with metabolites of toxic chemicals and drugs is presented in Chapter 2.

Another enzyme of importance in the antioxidant defense against quinone compounds is quinone reductase, also known as DT-diaphorase or Nmo-1 (Nebert et al. 1990). Quinones can undergo either one-electron reduction to a semiquinone or two-electron reduction to a hydroquinone. Semiquinones can be produced by one-electron reduction from a variety of electron donors, such as flavoproteins, and can react with oxygen to produce superoxide radical to regenerate the starting quinone:

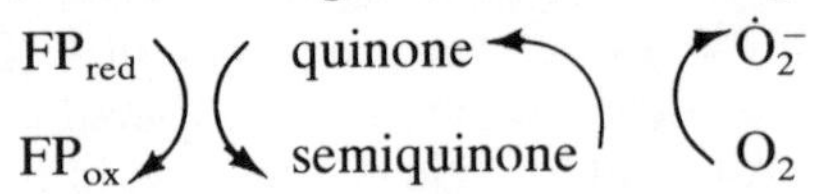

There is thus a potential for futile redox cycling in quinone metabolism that leads to increased formation of superoxide (discussed further in Chapter 2). This cycling can be circumvented by the two-electron transfer from NADH or NADPH catalyzed by quinone reductase that results in production of hydroquinones, which are usually less reactive and more easily conjugated and excreted:

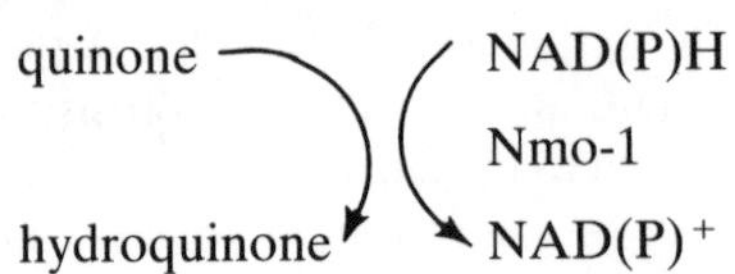

Hence, Nmo-1 may protect cells from the toxic effects of quinones. In the mouse, Nmo-1 is part of the [Ah] gene battery and as such its expression in tissues can be regulated by exogenous effectors such as dioxin (2,3,7,8 tetrachlorodibenzo-p-dioxin) (TCDD). The regulation of expression of this gene in humans has not been characterized.

NONENZYMATIC ANTIOXIDANTS

In addition to enzymatic defense mechanisms, there is an extensive nonenzymatic antioxidant network composed of several different types of lipid-

and water-soluble small molecules that have in common the ability to scavenge free radicals. Free radicals have been defined in several different ways. For this discussion, any species with one or more unpaired electrons will be designated as a free radical. Under this definition, molecular oxygen and most transition metals are free radicals, as are superoxide anions and hydroxyl radicals; hydrogen peroxide is not. Free radicals tend to engage in reactions that eliminate their unpaired electron(s). This end can be accomplished by reaction with another free radical (for example, the non-enzymatic dismutation of superoxide) or by loss of the unpaired electron to a nonfree radical acceptor molecule in a one-electron reduction or by gain of an electron from a nonfree radical donor molecule in a one-electron oxidation. Because concentrations of free radicals are low under most conditions, reactions with nonfree radicals are far more likely to occur. If the free radical reacts with a nonfree radical (having no unpaired electrons), then, whether an electron is gained or lost, another free radical will be created. Thus free-radical reactions tend to propagate in chains that are sustained as long as the free radicals created in the chain reactions have sufficient reactivity to continue to cause one-electron oxidation or reduction. Free-radical chains can be terminated by reaction with another free radical or by reaction with an acceptor or donor molecule whose free-radical product is stable and insufficiently reactive to continue the cascade of one-electron transfers. Most free-radical scavengers are of this latter type.

Among the lipid-soluble, free-radical scavengers are β-carotene and vitamin E; among those soluble in water are ascorbic acid, uric acid, and GSH. Vitamin E or α-tocopherol reacts with a variety of free radicals at fairly high rates to produce a chromanoxyl free radical (VitE-$\dot{O}$) which is known to be stable and can react with itself or with ascorbate to regenerate native vitamin E (VitE-OH) (Cadenas 1989). Because of its lipid solubility, vitamin E is instrumental in the inhibition of lipid peroxidation in membranes. Ascorbic acid is a nonspecific radical scavenger, capable of reacting with potent oxidants such as hydroxyl radicals to produce a stable semiquinone free radical that can in turn reduce another oxidizing radical (which can also be $\dot{O}H$) to give fully oxidized dehydroascorbate. Ascorbate reacts with peroxyl radicals (RO$\dot{O}$) and singlet oxygen as well. Some of the reactions of glutathione have already been discussed in connection with glutathione peroxidase and H_2O_2. Glutathione interacts readily with free radicals as well, usually producing the thiyl radical (G$\dot{S}$) as the major product, and with singlet oxygen. Thiyl radicals are not without reactivity (see Chapter 2), and some evidence for inactivation of enzymes by sulphenyl-peroxo radicals (GSO$\dot{O}$), produced by the addition of oxygen to the thiyl radical, has been obtained (Cadenas 1989).

OXIDATIVE DAMAGE

The combined presence of molecular oxygen and transition metal ions in aerobic living systems creates the ongoing potential for generation of ROS.

Any condition in which the generation of ROS is increased or in which the antioxidant defense is diminished constitutes oxidative stress. Some conditions that may result in oxidative stress include: increased levels of transition metals or their reactive forms, such as occurs in hemochromatosis or tissue injury; depletion of the nonenzymatic antioxidant defense by dietary deficiency or by excess antioxidant consumption during the metabolism of drugs and xenobiotics (see Chapter 2); increased generation of ROS by granulocyte activation (see Chapter 3), ionizing radiation or futile metabolic cycling (see Chapter 2); the presence of genetic variants of antioxidant enzymes that are less functional than wild-type enzymes (see Chapter 7). Oxidative stress that is severe or persistent enough, regardless of source, brings with it the likelihood that some form of oxidative damage will occur. It is also important to remember that the antioxidant defense is not static and that oxidative stress is a potent inducer of many types of protective enzymes (see Chapters 5 and 6).

As previously mentioned, ROS, and hydroxyl radical in particular, are capable of reacting with virtually all types of biomolecules. Despite this multiplicity of possible reactants, two types of molecules, DNA and lipids, are generally regarded as the significant targets for oxidative damage. There is no question that proteins can also be affected, but because they are usually present in multiple copies and are constantly turning over in tissues, damage to individual protein molecules is rarely of consequence and, as noted, may even serve a sacrificial, protective function. An exception to this may be the oxidative modification of proteins in low-density lipoprotein, which permits uptake by the scavenger receptor (see Chapter 4). There may be other exceptions as well, such as postischemic damage to myocardial contractile proteins as a possible source of impaired contractility in heart attacks. Damage to DNA brings with it the possibility of mutations that, if occurring in germ cells, may be heritable, and if in somatic cells, may contribute to carcinogenesis. DNA damage is discussed in detail in Chapter 5. Oxidative damage to lipid is manifest mainly as peroxidation and subsequent chain scission with release of cytotoxic aldehydes. It is significant because lipid peroxidation may be propagated by chain reactions in membranes and can result in crosslinking reactions, in substantial alterations in membrane fluidity and in membrane structure, the end point of which may be cell death.

Lipid peroxidation is initiated by the attack of a sufficiently reactive species, such as hydroxyl radical, on an esterified or free fatty acid (Halliwell and Gutteridge 1990). A hydrogen atom is removed, creating a carbon-centered free radical that, in polyunsaturated fatty acids, may be stabilized by an intramolecular rearrangement to a conjugated diene. The conjugated diene rapidly reacts with oxygen to produce a hydroperoxy radical:

$$\text{Lipid-H} + \dot{\text{O}}\text{H} \rightarrow \text{Lipid} + H_2O$$

$$\text{Lipid} \rightarrow \text{Intramolecular rearrangement (conjugated diene)} + O_2 \rightarrow \text{Lipid-}\dot{\text{O}}_2$$

The hydroperoxy radical thus created is sufficiently reactive to abstract a hydrogen atom from another lipid molecule, creating another carbon-centered lipid radical so that a propagated chain reaction may ensue:

$$\text{Lipid-}\dot{O}_2 + \text{Lipid H} \rightarrow \text{Lipid-}O_2H + \text{Lipid}$$

The hydroperoxy radical is thereby converted to a lipid hydroperoxide, lipid-O_2H, which, although fairly stable under physiologic conditions, can undergo further reactions if transition metal ions are present in reactive forms. Many protein-bound forms of iron such as hemoglobin, peroxidase, and cytochromes are effective in these reactions, although transferrin is not. Reaction with ferrous compounds yields alkoxy radicals (Lipid-$\dot{O}$) whereas reaction with ferric compounds produces peroxyl radicals (Lipid-$\dot{O}_2$), both of which can propagate chain reactions by abstracting hydrogen from other fatty acids:

$$\text{Lipid-}O_2H + Fe^{++}\text{-complex} \rightarrow Fe^{+++}\text{ complex} + OH^- + \text{lipid-}\dot{O}$$

$$\text{Lipid-}O_2H + Fe^{+++}\text{-complex} \rightarrow \text{Lipid-}\dot{O}_2 + H^+ + Fe^{++}\text{-complex}$$

Thus, if lipid peroxidation is initiated in the presence of reactive iron compounds, a variety of radical species can be produced, including carbon-centered, alkoxy, and peroxy radicals, all of which are capable of further hydrogen abstraction and of propagated chain reactions. In addition to the preceding reactions, fatty acid peroxides also undergo chain scission to produce a variety of degradation products, such as alkanals, alkenals, and hydroxylated alkenals, some of which may be strongly reactive (Halliwell and Gutteridge 1990). Membranes in which extensive peroxidation has occurred have altered fluidity, decreased membrane potential, and increased permeability and may ultimately rupture (Halliwell and Gutteridge 1989). Some of the end products of lipid peroxide fragmentation, such as malondialdehyde, 4-hydroxynonenal, and 4,5-dihydroxyldecenal, exert a variety of effects on cells. They can inhibit macrophage action, crosslink proteins, inactivate enzymes, and act as chemotactic agents for phagocytes (see Chapter 4). These products can also escape from the membrane to spread disturbances throughout the cell and even outside the cell.

The nature of the initiating reactions in lipid peroxidation is not clear. Hydroxyl radicals are reactive enough to begin the process and are known to be produced in membrane systems, but addition of hydroxyl radical scavengers has little effect on the observed rates of peroxidation. Other candidates suggested as initiators are hydroperoxy radical, $H\dot{O}_2$, and ferryl radical $FeOH^{+++}$ (or FeO^{++}), in which the iron is in a 4+ oxidation state. However, the question of initiation may be somewhat moot because lipid hydroperoxides, lipid-O_2H, appear to be present in almost all unsaturated fatty acid mixtures; they may be formed by reactions between membrane lipids and singlet oxygen and are synthesized by at least two widely distributed membrane enzymes, cyclo-oxygenase and lipoxygenase, during the formation of prostaglandins and leukotrienes. Because lipid hydroperoxides can be decomposed by reactive iron compounds to reactive alkoxyradicals, lipid-$\dot{O}$, the propagated processes of peroxidation may actually begin with them.

BIOMEDICAL CONSEQUENCES OF OXIDATIVE DAMAGE

Because the potential for oxidative damage to biomolecules is latent in all aerobic cells and may be unmasked by oxidative stresses of many kinds, one may legitimately ask whether any pathologic conditions exist in which oxidative stress or damage or both, play a primary, causative role. Increased lipid peroxidation in animal tissues, for example, has been shown to occur by a variety of techniques in a number of disease states (for example, muscular dystrophy, multiple sclerosis) and toxicant exposures (Hunter and Mohamed 1986; Hunter et al. 1985; Comporti 1985). The observed increases in lipid peroxidation, however, may be secondary to tissue damage or cell death from other causes, with the lipid peroxidation occurring as a result of liberation of transition metal ions that occurs with cell damage. Thus, although the progression of a disease process may involve oxidative damage and the outcome of the process may be ameliorated by chelating agents or chain-breaking antioxidants, one cannot conclude that oxidative damage is a primary cause unless it can be shown that oxidative damage precedes cell damage and that its prevention by antioxidants or chelating agents or both prevents cell damage as well. The list of conditions in which the involvement of oxygen radicals has been suggested is long. For most of them the requisite proof of the primacy of oxidative damage in the disease process is either lacking or not rigorous.

Evidence for the role of oxidative stress or damage in particular disease processes generally derives from several complementary methods. Some of the approaches used are:

1. Demonstration that the disease process is inhibited or prevented by addition of SOD, catalase, or the nonenzymatic SOD surrogate, Cu-(DIPS), singly and in combination: Positive results imply that superoxide or hydrogen peroxide or both play a role in the process. Appropriate controls with heat-denatured protein or apoenzyme are required because inactive SOD protein may also act as singlet oxygen and hydroxyl radical scavenger (Matheson et al. 1975).
2. Exacerbation or mimicking of the process by addition of a superoxide generating system such as xanthine/xanthine oxidase (X/XO) or direct addition of hydrogen peroxide. However, X/XO generates $\dot{O}_2^-$ extracellularly, and $\dot{O}_2^-$ does not permeate most cells (although it can enter red blood cells through the anion channel) so that a negative result may be inconclusive.
3. Inhibition of the process by hydroxyl radical scavengers, thus implying the involvement of hydroxyl radicals. However, no scavenger is spe-

cific for the hydroxyl radical, and in any case it may be generated in a site-specific manner that is resistant to scavengers in the bulk phase.

4. Direct detection of the relevant free radical by electron spin resonance using appropriate spin traps or by detection of a unique pattern of reactants that only arises from a particular radical (for example, characteristic patterns of aromatic hydroxylations caused by hydroxyl radicals). When correctly applied, these techniques can yield convincing evidence that a particular radical is present but cannot directly prove involvement of the radical in the disease process in question.
5. Manipulation of gene expression through techniques of molecular biology and observation of the effects of over- and underexpression of particular genes in disease processes. Analyses of mutants in which particular genes are inactive have provided valuable insight into the role played by many genes in modulating oxidative stress (see Chapter 5). The use of expression vectors bearing cloned genes in cell lines provides complementary data on what happens when particular genes are overexpressed.
6. Modulation of the process by antioxidants. This approach has been widely employed in analyzing the role of oxidative damage in carcinogenesis and is discussed in detail in Chapter 8. It has been employed less extensively to study atherosclerosis (see Chapter 4).

This book focuses on those pathologic processes for which strong evidence exists that oxidative stress or damage or both play a primary role: carcinogenesis and two types of cardiovascular disease, atherogenesis and ischemia/reperfusion injury. This is not to suggest that oxidative damage does not play a role in other disease processes; it almost certainly does so because of its ubiquitous nature. As our knowledge of pathologic mechanisms increases, it is likely that our appreciation of the centrality of oxidative stress and damage to the metabolic processes of aerobic systems will grow.

Carcinogenesis is a multistage process, in which several crucial genetic alterations accumulate in a cell, the result being that normal growth control is lost. The process may be influenced by promoters that alter the expression of genes controlling cellular differentiation and that favor cell growth. Selective cell death may also play a role in the process by stimulating division in the surviving cell population, among which genetically altered cells may be included. Oxidative damage appears to participate in this process in several ways:

1. The crucial genetic changes involved in initiation appear to be mutations that result in inactivation of antioncogenes or activation of proto-oncogenes. Any agents capable of causing DNA damage can thus act as initiators. There is abundant evidence, discussed in detail in Chapter 5, that ROS can cause extensive DNA damage and are mutagenic. Ionizing radiation, which acts mainly by generating ROS, is both mutagenic and carcinogenic.

2. Tumor promotion has been examined in several experimental systems using specific tumor promoters, such as phorbol esters (for example, TPA). The actions of tumor promoters can be mimicked by reactive oxygen-generating systems (for example, X/XO): tumor promoters have been shown to provoke endogenous production of ROS in several cell lines, and tumor promotion can be blocked by SOD and by a variety of antioxidants. Thus, oxidative stress or damage is strongly implicated in tumor promotion. Chapter 8 provides a more complete discussion.
3. As previously noted, oxidative damage can cause cell death, both by lipid peroxidation and as a consequence of DNA damage. Because cell death results in compensatory cell division in the surviving population, it can favor the clonal expansion of initiated cells. The important role that increased cell division may play in carcinogenesis has been described in a recent mathematical model (Cohen and Ellwein 1990).
4. Oxidative stress or damage constitutes a powerful signal to cells that results in activation and increased expression of many genes. Among the induced genes may be antioxidant genes, such as SOD and Nmo-1, and protooncogenes, such as c-myc and c-fos. These altered patterns of gene expression may result in a shift towards cell proliferation, as opposed to cell differentiation, and may thus promote tumor growth. A detailed discussion of oxidative stress and damage as a regulator of gene expression is given in Chapter 6.
5. The central role of iron as a catalyst of oxidative stress and damage has already been described. A thorough discussion of the epidemiologic evidence for iron as a co-carcinogen is provided in Chapter 7.

Recognition of the importance of oxidative damage in atherogenesis is now a dominant concept in the field. There is abundant evidence that oxidative alteration of low-density lipoprotein can generate a variety of cytotoxins and chemoattractants that may be instrumental in atherogenesis and that oxidatively altered low-density lipoprotein may be the form taken up by scavenger receptors on intimal fibroblasts, resulting in their conversion to foam cells. A lucid, comprehensive exposition of this topic is presented in Chapter 4.

When the blood flow to a tissue is restricted (ischemia), there is progressive development of tissue damage due to the decreased delivery of oxygen and metabolic substrates (for example, glucose) and to the gradual accumulation of metabolic by-products. The re-establishment of the blood supply (reperfusion), which is necessary to prevent ischemic necrosis, should reverse these effects. There is, however, substantial evidence that a paradoxical acceleration of tissue injury occurs because of the generation of ROS. Changes occur in the ischemic tissue that, on reintroduction of oxygenated blood, lead to formation of ROS, granulocyte infiltration, and microvascular dysfunction. A detailed description and critical review of these phenomena are found in Chapter 3. An understanding of the mechanisms by which ischemia/reperfusion damage occurs may suggest strategies for its prevention, which may be important in controlling the effects of myocardial infarctions and brain and spinal cord injuries.

Present therapy for myocardial infarction includes the use of thrombolytic agents, such as streptokinase and tissue plasminogen activator, that act to produce rapid reperfusion of ischemic myocardial tissue. Evidence from animal studies suggests that reperfusion of the heart, even after brief ischemic periods, can result in prolonged depression of contractility and arrhythmias and that these effects can be mitigated by use of antioxidants (SOD, catalase), iron chelators (desferroxamine), or radical scavengers (mannitol). The application of these findings to humans awaits the performance of appropriate clinical trials.

In the brain and spinal cord, traumatic or ischemic injury provokes extensive tissue damage that may reflect an especially vigorous generation of ROS and subsequent lipid peroxidation. Neural tissue is rich in membrane lipids containing polyunsaturated fatty acids. It is relatively poor in antioxidant enzyme content, and some areas of the brain are rich in iron (globus pallidus, substantia nigra, circumventricular area). When they are injured, brain cells can readily release some of their iron into CSF, which lacks iron-binding capacity. In addition, as a consequence of a specific active transport system in the choroid plexus, the concentration of ascorbate is about 10 times higher than plasma levels; even higher levels can be found in neural cells (Spector and Eels 1984). The iron-ascorbate mixture released from injured neuronal tissue can catalyze a particularly vigorous course of lipid peroxidation, which can exacerbate tissue damage; ischemia and reperfusion can lead to a similar result. The recognition that iron-catalyzed lipid peroxidation may contribute to brain and spinal cord injury following trauma or ischemia led to the development of a group of hydrophobic steroid-based iron chelators (the lazaroids) that, in animal studies, have proved useful in diminishing both ischemia/reperfusion and post-trauma injuries (Hall 1988; Hall and Travis 1988; Hall et al. 1988).

With deepening knowledge of the processes underlying oxidative stress and damage, there has come a developing sense that it is one of the fundamental pathophysiologic mechanisms affecting aerobic living systems. The ramifications will become more and more apparent as experimental data accumulate.

REFERENCES

Aebi, H. E., and Wyss, S. R. (1978). Acatalasemia. In J. B. Stanbury, J. B. Wyngaarden, D. S. Fredrickson, and M. S. Brown, eds. *The Metabolic Basis of Inherited Disease.* New York: McGraw Hill. pp. 1792–1807.

Aruoma, O. I., Bomford, A., Polson, R. J., and Halliwell, B. (1988). Non-transferrin-bound iron in plasma from hemochromatosis patients: effect of phlebotomy therapy. *Blood*, 72:1416–19.

Ashton, N. (1979). The pathogenesis of retrolental fibroplasia. *Ophthalmology*, (Rochester) 86:695–99.

Balentine, J. D. (1978). Experimental pathology of oxygen toxicity. In F. F. Jobsis, ed. *Oxygen and Physiological Function.* Dallas, Texas: Professional Information Library. pp. 331–44.

Barnes, S. D., Agee, C. C., Peace, R. J., and Leffler, C. W. (1983). Effects of elevated PO_2 on tracheal explants. *Respir. Physiol.*, 53:285–93.

Betts, E. K., Downes, J. J., Schaffer, D. B., and Johns, R. (1977). Retrolental fibroplasia and oxygen administration during general anesthesia. *Anesthesiology*, 47:518–20.

Bleijenberg, B. J., van Eikj, H. G., and Leijnse, B. (1971). The determination of non-heme iron and transferrin in cerebrospinal fluid. *Clin. Chim. Acta.*, 31:277–81.

Bonikos, D. S., Bensch, K. G., and Northway, W. H., Jr. (1976). Oxygen toxicity in the newborn. The effect of chronic continuous 100% oxygen exposure on the lungs of newborn mice. *Am. J. Pathol.*, 85:623–50.

Boveris, A., and Cadenas, E. (1982). In L. W. Oberley, ed. *Superoxide Dismutase.* vol 2. Boca Raton, FL: CRC Press. pp. 15–30.

Cadenas, E. (1989). Biochemistry of oxygen toxicity. *Ann. Rev. Biochem*, 58:79–110.

Cohen, S. M., and Ellwein, L. B. (1990). Cell proliferation in carcinogenesis. *Science*, 249:1007–11.

Comporti, M. (1985). Lipid peroxidation and cellular damage in toxic liver injury. *Lab. Invest.*, 53:599–23.

Davis, W. B., Rennard, S. I., Bitterman, P. B., and Crystal, R. G. (1983). Pulmonary oxygen toxicity: early reversible changes in human alveolar structures induced by hyperoxia. *N. Engl. J. Med.*, 309:878–83.

Deneke, S. M., and Fanburg, B. I. (1982). Oxygen toxicity of the lung: an update. *Br. J. Anaesth.*, 54:737–49.

DiGuiseppi, J., and Fridovich, I. (1983). The toxicology of molecular oxygen. *Crit. Rev. Toxicol.*, 12:315–42.

Fenn, W. O., Gerschman, R., Gilbert, D. L., Terwillinger, D. E., and Cothran, F. V. (1957). Mutagenic effects of high oxygen tensions on E. coli. *Proc. Natl. Acad. Sci. USA*, 43:1027–31.

Forstrom, J. W., Zakowski, J. J., and Tappel, A. L. (1978). Identification of the catalytic site of rat liver glutathione peroxidase as selenocysteine. *Biochemistry*. 17:2639–44.

Fridovich, I. (1974). Superoxide dismutases. *Advances in Enzymology*, 41:35–48.

Fridovich, I. (1975). Superoxide dismutases. *Ann. Rev. Biochem.*, 44:147–59.

Fridovich, I. (1978). The biology of oxygen radicals. *Science*, 201:875–80.

Fridovich, I., and Freeman, B. (1986). Antioxidant defenses in the lung. *Ann. Rev. Physiol.*, 1986:693–702.

Frimer, A. A. (1982). In L. W. Oberley, ed. *Superoxide Dismutase*, Boca Raton, FL: CRC Press. pp. 83–125.

Gregory, E. M., Goscin, S. A., and Fridovich, I. (1974). Superoxide dismutase and oxygen toxicity in a eukaryote. *J. Bacteriol*, 117:456–60.

Gutteridge, J. M. C. (1986). Iron promoters of the Fenton reaction and lipid peroxidation can be released from hemoglobin by peroxides. *FEBS Lett.*, 20:291–95.

Gutteridge, J. M. C., Rowley, D. A., and Halliwell, B. (1981). Superoxide-dependent formation of hydroxyl radicals in the presence of iron salts. Detection of "free" iron in biochemical systems by using bleomycin-dependent degradation of DNA. *Biochem. J.*, 199:263–65.

Hall, E. D. (1988). Effects of the 21-aminosteroid U74006F on post-traumatic spinal cord ischemia in cats. *J. Neurosurg.*, 68:462–65.

Hall, E. D., and Travis, M. A. (1988). Inhibition of arachidonic acid-induced vasogenic brain edema by the non-glucocorticoid 21-aminosteroid U74006F. *Brain Res.*, 451:350–52.

Hall, E. D., Yonkers, P. A., McCall, J. M., and Braughler, J. M. (1988). Effects of the 21-amino-steroid U74006F on experimental head injury in mice. *J. Neurobiol.*, 68:456–61.

Halliwell, B. (1984). *Chloroplast Metabolism: The Structure and Functions of Chloroplasts in Green Leaf Cells.* Oxford: Oxford University Press (Clarendon).

Halliwell, B., and Gutteridge, J. M. C. (1984). Oxygen toxicity, oxygen radicals, transition metals and disease. *Biochem. J.*, 219:1–14.

Halliwell, B., and Gutteridge, J. M. C. (1989). *Free Radicals in Biology and Medicine.* 2nd ed. Oxford: Oxford University Press (Clarendon).

Halliwell, B., and Gutteridge, J. M. C. (1990). Role of free radicals and catalytic metal ions in human disease: an overview. In L. Packer and A. N. Glazer, eds. *Methods in Enzymology.* vol. 186. pp. 1–85.

Hunter, M. I. S., and Mohamed, J. B. (1986). Plasma antioxidants and lipid peroxidation products in Duchenne muscular dystrophy. *Clin. Chem. Acta.*, 155:123–31.

Hunter, M. I. S., Niemadin, B. C., and Davison, D. L. W. (1985). Lipid peroxidation products and antioxidant proteins in plasma and cerebrospinal fluid from multiple sclerosis patients. *Neurochem. Res.*, 10:1645–52.

Itabe, H., Kobayashi, T., and Inowe, K. (1988). Generation of toxic phospholipid(s) during oxyhemoglobin-induced peroxidation of phosphatidylcholines. *Biochim. Biophys. Acta*, 961:13–21.

Kaye, D. (1967). Effect of hyperbaric oxygen on aerobic bacteria in vitro and in vivo. *Proc. Soc. Exp. Biol. Med.*, 124:1090–93.

Kehrer, H. P., and Autor, A. P. (1978). The effect of dietary fatty acids on the composition of adult rat lung lipids: relationship to oxygen toxicity. *Toxicol. Appl. Pharmacol.*, 44:423–30.

Matheson, E. B. C., Etheridge, R. D., Kratowich, N. R., and Lee, J. (1975). The quenching of singlet oxygen by amino acids and proteins. *Photochem. Photobiol.*, 21:165–71.

Miller, J. W., and Winter, P. M. (1981). Clinical manifestations of pulmonary oxygen toxicity. *Int. Anesthesiol. Clin.*, 19:179–99.

Morris, E. R. (1983). An overview of current information on bioavailability of dietary iron to humans. *Federation Proceedings*, 42:1716–20.

Motten, A. G., and Chignell, C. F. (1983). Spectroscopic studies of cutaneous photosensitizing agents. III. Spin trapping of photolysis products from sulfanilamide analogues. *Photochem. Photobiol.*, 37:17–26.

Nebert, D. W., Petersen, D. D., and Fornace, A. J. (1990). Cellular responses to oxidative stress. The [AH] gene battery as a paradigm. *Environ. Health Perspect.*, 88:13–25.

Pulich, W. M., Jr. (1974). Resistance to high oxygen tension, streptonigrin and ultraviolet irradiation in the green alga chlorella sorokiniana strain ORS. *J. Cell Biol.*, 62:904–07.

Puppo, A. and Halliwell, B. (1988). Formation of hydroxyl radicals from hydrogen peroxide in the presence of iron. Is hemoglobin a biological Fenton reagent? *BioChem J.*, 249:185–90.

Rabinowitch, H., and Fridovich I. (1983). Superoxide radicals, superoxide dismutases and oxygen toxicity in plants. *Photochem. Photobiol.*, 37: 679–83.

Smith, T. C., Gross, J. B., and Wollman, H. (1985). The therapeutic gases; oxygen, carbon dioxide, helium and water vapor. In A. G. Gilman, L. S. Goodman, T. W. Rall, and F. Murad, eds. *Goodman & Gilman's The Pharmacological Basis of Therapeutics*. 7th ed. New York: Macmillan, pp. 322–28.

Spector, R., and Eels, J. (1984). Deoxynucleoside and vitamin transport into the central nervous system. *Federation Proceedings*, 43:196–200.

Staal, G. E. J., Vissero, J., and Veeger, C. (1969). Purification and properties of glutathione reductase of human erythrocytes. *Biochim. Biophys. Acta*, 185:39–48.

Taylor, D. W. (1954). Effects of high oxygen pressures on adrenalectomized, treated and untreated rats. *J. Physiol.*, 125:46P.

Tierney, D. F., Ayers, L., and Kasuyama, R. S. (1977). Altered sensitivity to oxygen toxicity. *Am. Rev. Respir. Dis.*, 115(6):59–65.

Turrens, J. F., Crapo, J. D., and Freeman, B. A. (1984). Protection against oxygen toxicity by intravenous injection of liposome-entrapped catalase and superoxide dismutase. *J. Clin. Invest.*, 73:87–95.

Valentine, J. S., Miksztal, A. R., and Sawyer, D. T. (1984). Methods for the study of superoxide chemistry in nonaqueous solutions. *Methods Enzymol.*, 105:71–81.

Weir, M. P., Gibson, J. F., and Peters, T. J. (1984) Hemosiderin and tissue damage. *Cell Biochem. Funct.*, 2:186–94.

Zigler, J. S., and Goosey, J. D. (1981). Photosensitized oxidation in the ocular lens: evidence for photosensitizers endogenous to the human lens. *Photochem. Photobiol.*, 33:869–74.

2

Free Radical Metabolites of Toxic Chemicals and Drugs as Sources of Oxidative Stress

RONALD P. MASON

For a variety of reasons, the possibility of free radical metabolism has not received much attention in the past, although Michaelis of the Michaelis-Menten equation was interested in free radical metabolites and their importance in biochemistry in the 1930s. One reason for the late development of this area is that most biochemicals, as opposed to aromatic drugs and industrial chemicals, are not easily metabolized through free radical intermediates. Nonetheless, it is now clear that in some cases the free radical metabolites of drugs and toxic chemicals, or the superoxide, alkoxyl, and peroxyl radicals derived from them, contribute significantly to oxidative stress. Reactions of these free radicals may be important causes of damage to biochemical macromolecules and of toxicity. In this chapter we examine some of the mechanisms by which free radical metabolites of xenobiotics are created and how they interact with biologic systems.

A free radical is any organic molecule with an odd number of electrons. Even a simple organic molecule such as benzene can be transformed into three chemically distinct, highly reactive free radicals (Figure 2-1). One-electron oxidation, the removal of an electron from the pi-electrons, results in the formation of the benzene cation radical. One-electron reduction of benzene, the addition of an electron, results in the formation of the benzene anion radical. The third free radical is formed by the homolytic cleavage of one of the C-H bonds by ultraviolet light or other radiation to form a hydrogen atom and the phenyl radical.

Severe chemical conditions are necessary to form free radicals from benzene, but this is not the case for most aromatic compounds. In fact, many classes of free radicals are formed as a result of the metabolism of

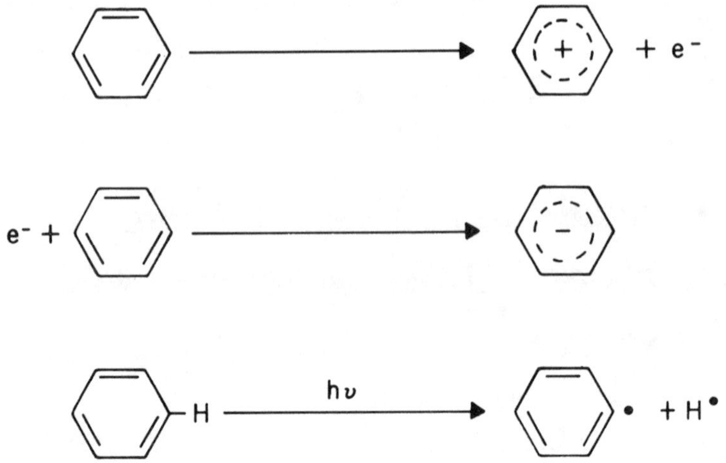

Figure 2-1 Free radicals from benzene.

chemicals. Of the three types of free radical metabolites, only radical anion metabolites participate in redox cycling, which results in the catalytic formation of superoxide. These species are analogous to the benzene anion (see Figure 2-1). They are formed by a one-electron transfer from an enzyme to an aromatic organic chemical, which may be either a drug or an industrial chemical. The futile metabolism to radical anions is the best known xenobiotic source of superoxide. Investigations of bipyridylium, azo, quinone, and nitro radical anion metabolite formation have been extensive (Mason 1979, 1982; Mason and Chignell 1981; Kappus and Sies 1981).

Other sources of superoxide and other oxygen-centered free radicals exist, but are not so widely known. Some types of xenobiotic free radical metabolites react with oxygen to form superoxide; however the reaction does not regenerate the parent compound, as in futile metabolism, but forms a stable metabolite of the compound.

In other cases, xenobiotic free radical metabolites react with oxygen, not by electron transfer to form superoxide, but by covalent bond formation to form peroxyl free radicals. Lastly, organic hydroperoxides are decomposed by many hemoproteins to free radicals.

FUTILE METABOLISM AS A SOURCE OF SUPEROXIDE

Paraquat and Other Bipyridylium Compounds

The herbicide paraquat and related bipyridylium dications such as diquat can undergo a one-electron reduction to form very stable free radicals. Michaelis and Hill (1933) showed that the paraquat free radical can utilize

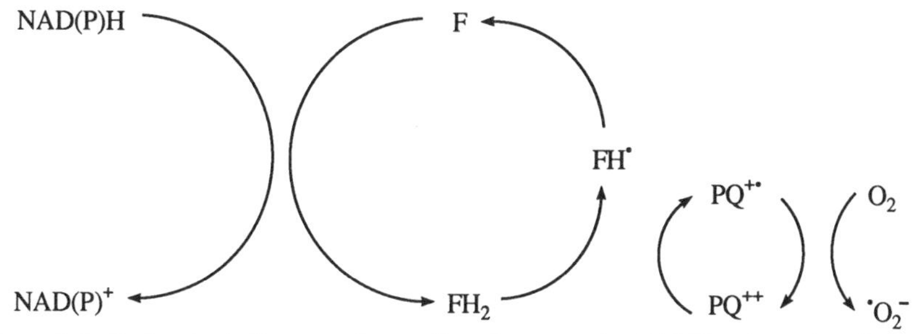

Figure 2-2 Futile metabolism or redox cycling of paraquat by NAD(P)H-dependent flavoenzymes.

molecular oxygen as a one-electron acceptor to form the superoxide anion radical with the regeneration of the paraquat dication.

$$P\dot{Q}^{+} + O_2 \rightarrow PQ^{++} + \dot{O}_2^{-}$$

Homer and colleagues (1960) proposed that the reduction of paraquat to its free radical was an essential step in its herbicidal mode of action because a correlation was found between the reduction potential of paraquat analogs and their herbicidal activity. Paraquat is reduced to its free radical within chloroplasts during photosynthesis, and the herbicidal activity of paraquat requires light for electron transport. Plant leaves incubated in paraquat solutions accumulated malondialdehyde, indicating that lipid peroxidation occurs (Dodge 1971). This lipid peroxidation is thought to be mediated by the one-electron reduction of paraquat and the subsequent transfer of the electron to molecular oxygen, resulting in superoxide formation (Harbour and Bolton 1975).

For over 20 years, paraquat has been known to be reduced in anaerobic microsomal incubations to a free radical as evidenced by its visible absorption spectrum (Gage 1968). This free radical is a deep blue. The electron spin resonance (ESR) spectrum is, however, a better means of identifying free radicals because, like nuclear magnetic resonance, ESR is much more specific than ultraviolet-visible spectroscopy. Paraquat serves as an ideal model compound for investigating free radical-mediated toxicity because it has no known metabolism other than the free radical metabolism.

In microsomal systems, the enzymatic reduction of paraquat to its cation radical is catalyzed by the flavoenzyme, NADPH-cytochrome P-450 reductase. The paraquat radical is stable in the absence of oxygen. In the presence of oxygen, paraquat is re-formed, and superoxide is generated in a catalytic fashion with no net change occurring to the paraquat molecule (Figure 2-2). This process has been termed futile metabolism (Mason 1979, 1982) or redox cycling (Kappus and Sies 1981). The mechanism of paraquat poisoning in man and other mammals is generally thought to be a superoxide-mediated toxicity that is completely analogous to the herbicide mode of action. The lung is the site of injury by paraquat because it accumulates there (Rose and Smith 1977). The energy-dependent uptake of paraquat and the subsequent free radical formation are cell-specific. Paraquat free radical formation occurs with clara cells and alveolar type 2 cells, but not with alveolar macrophages (Horton et al. 1986). Diquat, morfamquat, and other bipyridylium compounds do not affect the lung as seriously, but these compounds do cause liver damage. We have shown that diquat, paraquat, benzyl viologen, and morfamquat are reduced by rat hepatocytes to their respective radical cations (DeGray et al. 1991).

Quinones

The quinone moiety is found in pigments isolated from a variety of plants and fungi, some of which are clinically important antitumor drugs (Powis 1989; Sinha 1989). Although menadione (vitamin K_3) is used therapeutically, it is also cytotoxic and causes a marked decrease of intracellular thiols such as glutathione, the formation of superoxide by futile cycling, the concomitant oxidation of reduced pyridine nucleotides, alterations in intracellular calcium ion homeostasis, and the death of isolated hepatocytes (Gant et al. 1988; Rao et al. 1988b and references therein).

Doxorubicin, daunorubicin, and other anthracycline anticancer drugs are known to be carcinogenic, mutagenic, and cardiotoxic (Powis 1989; Sinha 1989). The first evidence of enzymatic semiquinone formation from a quinone anticancer drug was indirect. Handa and Sato (1975) demonstrated that daunorubicin and doxorubicin mediated the formation of superoxide in microsomal incubations containing NADPH. Later, they also demonstrated that these compounds stimulated aerobic NADPH oxidation in the absence of any net reduction of these antitumor compounds (Handa and Sato 1976). The presence of semiquinone metabolites of anthracyclines has been demonstrated with ESR in anaerobic incubations containing microsomes, purified NADPH-cytochrome P-450 reductase, and even in incubations of tumor cells. Analysis of the high-resolution ESR spectrum of the enzymatically generated daunorubicin semiquinone has been reported (Schreiber et al. 1987; Jülich et al. 1988).

Semiquinone involvement in redox cycling with subsequent biochemical and toxicologic consequences is the subject of many articles in Volume 8 of *Free Radical Research Communications*, 197–415 (1990). There has also been recent interest in the redox cycling of the quinones formed from diethylstilbestrol and the catechols of estradiol (Liehr and Roy 1990).

Azo Compounds

Although red dye #2 is only a weak carcinogen, this compound was banned as a food dye by the Food and Drug Administration because of the risk resulting from its high consumption. The reductive metabolism of azo compounds such as red dye #2 by a wide variety of biologic systems has long been known. Sulfonazo III is a diazonaphthol compound used in the titrimetric determination of sulfates and organic sulfur. Sulfonazo III is

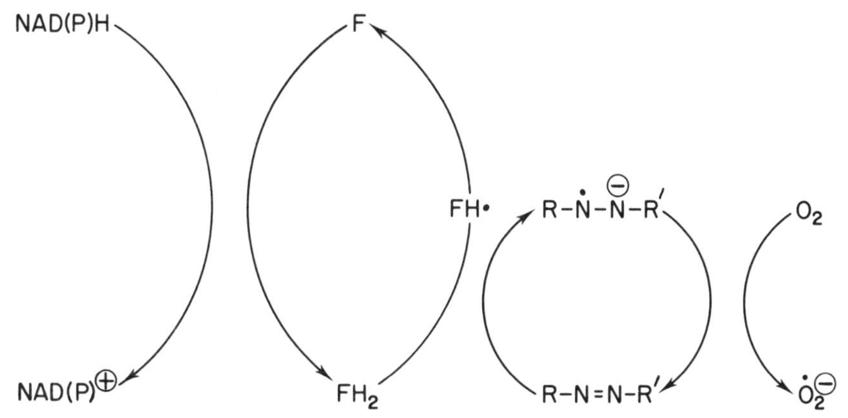

Figure 2-3 Futile metabolism or redox cycling of some azo compounds by NAD(P)H-dependent flavoenzymes.

structurally related to the monoazo food dyes such as red dye #2. We detected the ESR spectrum of a free radical metabolite of sulfonazo III in anaerobic rat hepatic microsomal incubations containing this azo dye and NADPH (Mason et al. 1977). The spectrum of the sulfonazo III free radical is characterized by a partially resolved 17-line hyperfine pattern and a g-value of the center line equal to 2.0034. Again, the scheme of futile metabolism (Figure 2-3) emphasizes the rapid air oxidation of the azo anion radical as the pivotal event (Mason et al. 1978). In such a scheme, there would be no net reduction of the azo compound because the parent compound would be re-formed. Sulfonazo III would thereby catalyze the production of superoxide anion radical, oxygen consumption, and NAD(P)H depletion.

The simplest method for detecting superoxide is the addition of superoxide dismutase to the reaction medium. This enzyme catalyzes the disproportionation of the superoxide anion radical to give back half of the superoxide as oxygen and reduces the other half to hydrogen peroxide: $2\dot{O}_2^- + 2H^+ \rightarrow O_2 + H_2O_2$. In such a reaction, the hydrogen peroxide formed by the disproportionation of the superoxide anion radical can itself be disproportioned by catalase to give back half of the oxygen as molecular oxygen: $2H_2O_2 \rightarrow O_2 + 2H_2O$. When both superoxide dismutase and catalase are added to the incubation, water is the only reduced species of oxygen that can accumulate.

In view of these considerations, the stimulation of oxygen uptake by sulfonazo III and the reversal of this stimulation by superoxide dismutase and catalase strongly indicate that the azo anion radical is formed in the presence of oxygen (Figure 2-4). When we examined the effect of 50 μM sulfonazo III on the NADPH-supported oxygen consumption by rat hepatic microsomes, we found that, indeed, the rate of oxygen uptake was increased tenfold over the basal rate and that this stimulation was partially reversed by superoxide dismutase. The presence of the superoxide anion

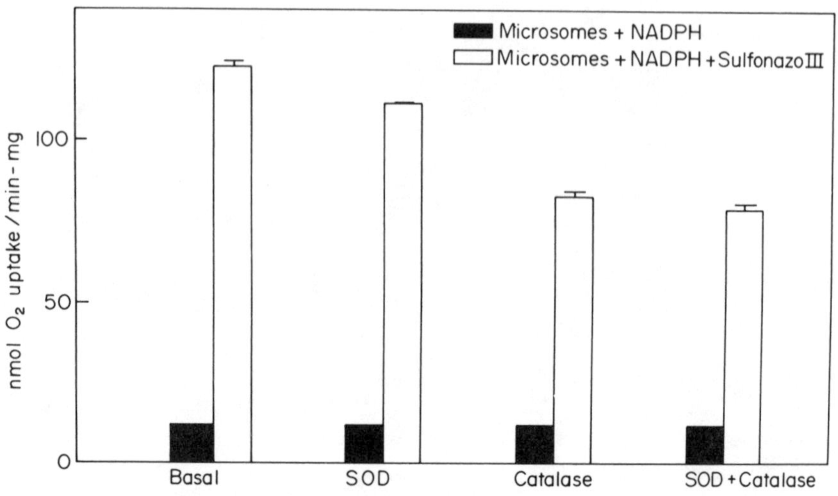

Figure 2-4 Effect of superoxide dismutase and catalase on sulfonazo III stimulation of oxygen consumption by rat hepatic microsomal incubations. (Data from Mason et al., 1978.)

radical strongly suggests that the sulfonazo anion free radical is formed by a microsomal reductase under aerobic conditions. The rate of oxygen uptake is over five times greater than that observed during normal cytochrome P-450-catalyzed reactions. As expected, the disproportionation of hydrogen peroxide by catalase also decreased the sulfonazo III-stimulated uptake of oxygen. The rate of dye disappearance in these incubations is only 2% of the rate of oxygen consumption, indicating that the consumption of oxygen is indeed catalytic. The oxidation of NADPH by microsomal incubations is also greatly increased by sulfonazo III, but it is not influenced by superoxide dismutase or catalase (Mason et al. 1978).

The Ca^{2+} indicator arsenazo III is closely related to sulfonazo III. At the concentrations used when it acts as a Ca^{2+} indicator, arsenazo III undergoes a one-electron reduction by rat liver microsomes (Docampo et al. 1983), mitochondria (Moreno et al. 1984), and cytosol (Moreno et al. 1985) to produce an azo anion radical as demonstrated by ESR spectroscopy. Either NADH or NADPH can serve as a source of reducing equivalents for the production of this free radical. Under aerobic conditions, addition of arsenazo III to rat liver microsomes produces an increase in electron flow from NAD(P)H to molecular oxygen, generating both superoxide anion and hydrogen peroxide (see Figure 2-3). The steady-state concentration of the azo anion radical, but neither oxygen consumption nor superoxide anion formation, is enhanced by calcium and magnesium, suggesting an enhanced azo anion radical-stabilization by complexation with the metal ions (Moreno et al. 1984). Accordingly, the arsenazo III anion radical ESR spectrum is abolished in the presence of paramagnetic metal ions (Gd^{3+} and Ni^{2+}) and enhanced in the presence of other diamagnetic metal ions (La^{3+}).

Compound	Structure
Nitrofurazone	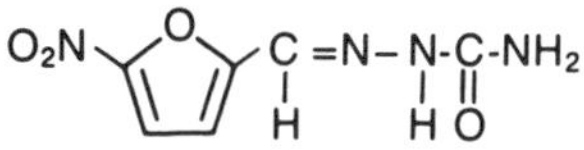 O_2N, O, C =N– N-C-NH$_2$, H, H O
Nitrofurantoin	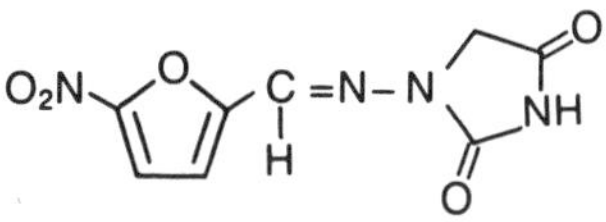 O_2N, O, C =N– N, O, NH, H, O
Benznidazole	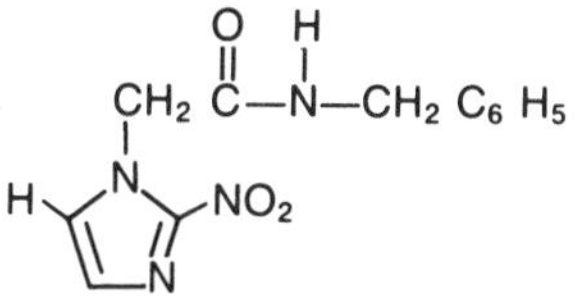 O H, CH_2 C—N—CH_2 C_6 H_5, H, N, NO_2, N
p-Nitrobenzoate	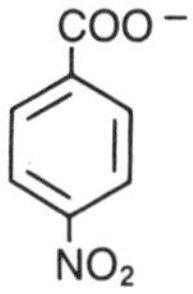 COO^-, NO_2
Metronidazole	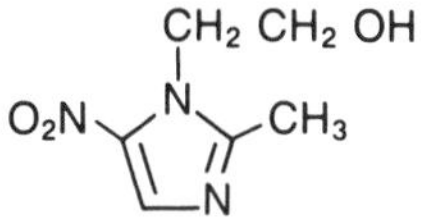 CH_2 CH_2 OH, O_2N, N, CH_3, N

Figure 2-5 Nitro drugs and model substrates.

One final point is that azo anion radical formation in the presence of oxygen does not lead to the formation of the ultimate products of azoreduction, aromatic amines and hydrazines, unless the oxygen concentration is low. The absence of these reductive metabolites in vitro or in vivo does not imply that azo reduction to the anion radical has not occurred. For instance, our microsomal oxygen uptake results imply that every molecule of sulfonazo III in the incubation is reduced to the anion radical once every 9 s (Mason et al. 1978), but the nearly quantitative air oxidation of the azo anion radical results in little net disappearance of the sulfonazo III until all of the oxygen is consumed. Therefore, better known methods of detecting drug metabolites such as high performance liquid chromatography (HPLC), which of course cannot detect unstable free radicals, may lead to erroneous conclusions concerning the extent of azo reduction.

Nitroaromatic Compounds

Nitroaryl and nitroheterocyclic compounds have enjoyed widespread use in medicine as antibiotics (Figure 2-5). The most widely employed topical substituted 5-nitrofuran, nitrofurazone, has been used as a food preservative, in treatment of patients with second- and third-degree burns, and as an antibacterial agent for the treatment or prevention of a wide variety of infections of the genitourinary tract. Nitrofurantoin is the substituted 5-nitrofuran administered most frequently for systemic infections, particularly those involving the urinary tract. Benznidazole is used as an antiprotozoal (Moreno et al. 1982), and metronidazole has been widely used for many years in the treatment of infections of *Trichomonas vaginalis*, amoebas, and girardia, and a host of anaerobic bacterial infections.

In our early investigation of the mechanism of rat hepatic mitochondrial and microsomal nitroreductase (Mason and Holtzman 1975a), we reported ESR and kinetic evidence that demonstrated that the first step in these nitroreductase reactions is the transfer of a single electron to nitro compounds to give the corresponding nitro anion free radical. For instance, in the case of nitrofurantoin, the interaction of the free electron with the nitrogens and protons gives a complex hyperfine pattern that has been analyzed and demonstrates that the free radical is simply nitrofurantoin plus an extra electron (Rao et al. 1987, 1988a).

When we examined the effect of 100 μM nitrofurantoin on the NADPH-supported oxygen consumption by hepatic or pulmonary microsomes, we found that, indeed, the rate of oxygen uptake was increased sevenfold over the basal rate and that this stimulation was partially reversed by superoxide dismutase. The presence of superoxide anion radical strongly suggested that the nitrofurantoin anion free radical is formed by microsomal nitroreductase under aerobic conditions (Mason and Holtzman 1975b). As expected, the disproportionation of hydrogen peroxide by catalase also decreased the nitrofurantoin-stimulated oxygen uptake. When both superoxide dismutase and catalase were added to the incubations, the nitrofurantoin-catalyzed oxygen consumption was decreased by over one third. We examined the paraquat-stimulated uptake of oxygen by microsomes to compare it with the nitrofurantoin-stimulated uptake. Paraquat stimulates the uptake of oxygen by microsomes less than an equal concentration of nitrofurantoin. Otherwise, the effect of superoxide dismutase or catalase is similar to that observed with the nitrofurantoin-stimulated oxygen uptake (Mason and Holtzman 1975b).

Our work on the effect of superoxide dismutase and catalase on the nitro compound-stimulated oxygen consumption by microsomes is consistent with the formation of nitroaromatic anion radicals under aerobic conditions and with the rapid air oxidation of these radical intermediates, which results in the catalytic generation of superoxide and the well-known oxygen inhibition of nitroreductases. We propose that the nitrofurantoin-catalyzed reduction of oxygen to superoxide and the hydrogen peroxide that forms from this superoxide may be responsible for some of the toxic manifestations that occur during nitrofurantoin therapy (Mason and Holtzman 1975b). For instance, we noted that the effects of the occasional cases of pulmonary edema and fibrosis caused by nitrofurantoin therapy are similar to the effects of paraquat poisoning. Subsequent work with animal models supported our proposal (Peterson et al. 1982; Boyd et al. 1979). Boyd and coworkers (1979) have shown that the acute toxicity of nitrofurantoin is markedly increased by vitamin E deficiency, an oxygen-enriched atmosphere, or a diet high in highly polyunsaturated fats. Further studies by Peterson and colleagues (1982) showed that decreasing the activity of selenium-dependent glutathinone peroxidase in 8-day-old chicks enhanced the acute toxicity of nitrofurantoin.

Carbon-Centered Free Radicals

Under nitrogen, the triarylmethane dye gentian violet undergoes a one-electron reduction by cytochrome P-450 to produce a carbon-centered free radical (Harrelson and Mason 1982). This free radical cannot be detected in the presence of oxygen, but the formation of superoxide can be detected (Fischer et al. 1984). The gentian violet radical apparently is formed under aerobic conditions, but it is oxidized by oxygen, which regenerates the dye and forms superoxide.

NONCATALYTIC, BUT STILL ENZYMATIC SOURCES OF SUPEROXIDE

Hydrazine Free Radicals

Phenylhydrazine is typical of a wide range of drugs that react with oxyhemoglobin by a redox mechanism in which oxyhemoglobin is oxidized to methemoglobin, and the drug is oxidized to a free radical (Smith and Maples 1985).

$$Hb(Fe^{2+})O_2 + PhNHNH_2 + H^+ \rightarrow Hb(Fe^{3+}) + Ph\dot{N} - NH_2 + H_2O_2$$

Individuals who have a glucose-6-phosphate dehydrogenase deficiency are especially susceptible to drug-induced hemolytic anemia, which results from this reaction. The oxidation of oxyhemoglobin to methemoglobin and to other hemoglobin-derived species, the formation of drug free radicals, and the concomitant generation of hydrogen peroxide have all been proposed to be involved in drug-induced hemolytic anemia.

In the presence of either oxyhemoglobin or methemoglobin, phenylhydrazine is oxidized to a species capable of the univalent reduction of molecular oxygen to superoxide (Goldberg et al. 1976; Misra and Fridovich 1976). Goldberg and colleagues (1976) have proposed this species to be phenyldiazine, whereas Misra and Fridovich (1976) have proposed that the phenylhydrazyl radical is the superoxide-forming intermediate. Note that the phenylhydrazyl radical is the conjugate acid of an anion free radical.

$$Ph\dot{N} - NH_2 \rightarrow H^+ + Ph\dot{N} - NH^-$$

Although the origin of superoxide is still in dispute, it is generally agreed that superoxide is derived from free molecular oxygen and not from the oxygen of oxyhemoglobin (Goldberg et al. 1976; French et al. 1978). In

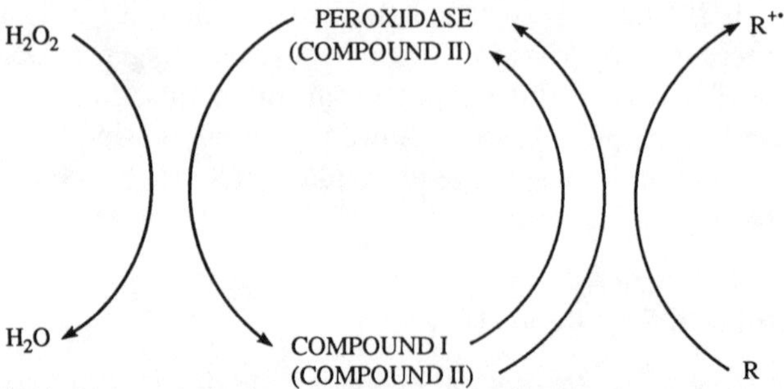

Figure 2-6 The formation of a wide variety of free radical metabolites by many peroxidases.

addition, the formation of superoxide in this reaction has been confirmed directly with ESR (Goldberg et al. 1979).

Superoxide dismutase is reported to inhibit phenylhydrazine-induced red blood cell hemolysis and peroxidation (Valenzuela et al. 1977). In apparent contradiction, superoxide dismutase is also reported not to inhibit either phenylhydrazine-induced hemolysis or oxyhemoglobin destruction (Goldberg and Stern 1977). In any case, in vivo lipid peroxidation does occur after treatment with phenylhydrazine in both circulating (Jain and Subrahmanyam 1978; Jain and Hochstein 1979) and splenic red blood cells (Jain and Subrahmanyam 1978).

Sulfur-centered Free Radicals

A wide variety of primarily aromatic compounds such as phenols are oxidized by horseradish peroxidase (HRP) to free radical metabolites in a reaction that requires hydrogen peroxide (Figure 2-6). Evidence that thiyl free radicals can thus be generated was found in the reports by Olsen and Davis (1976) of compound II formation in incubations of L-cysteine.

$$\text{HRP} + H_2O_2 \rightarrow \text{HRP-compound I(green)} + H_2O$$

$$\text{HRP-compound I} + \text{GSH } (\text{GS}^-) \rightarrow \text{HRP-compound II (red)} + \text{G}\dot{\text{S}}$$

$$\text{HRP-compound II} + \text{GSH } (\text{GS}^-) \rightarrow \text{HRP} + \text{G}\dot{\text{S}}$$

The formation of compound II, the second peroxidase intermediate, requires a one-electron transfer and is thus a good inferential indication that the thiyl free radical is being generated, but our ESR investigations are the first to provide direct evidence of enzymatic thiyl free radical formation. The spin-trapping technique was used because thiyl free radicals themselves cannot be detected in solution with ESR for theoretical reasons.

Because HRP is of plant origin, some may question the pharmacologic relevance of work done with this enzyme. On the other hand, saliva, tears, and milk contain lactoperoxidase, which is thought to have an antimicrobial function in vivo. This mammalian peroxidase has a soret optical spectrum from its heme group that is similar to that of thyroid peroxidase, intestine peroxidase, uterus peroxidase, eosinophil peroxidase, and prostaglandin H synthase. As such, lactoperoxidase appears to be a useful prototype for most mammalian hemoprotein peroxidases. The ESR spectrum of the spin-trapped cysteine thiyl radical is obtained with 0.1 mg/ml lactoperoxidase and 10 mM cysteine (Mottley et al. 1987). The structure of this radical adduct has been confirmed with mass spectrometry. The signal is totally inhibited by catalase and requires native lactoperoxidase but does not require added hydrogen peroxide. Under nitrogen, the thiyl radical formation requires hydrogen peroxide, and residual activity is inhibited by catalase.

We have investigated the consumption of oxygen because free radicals often react with oxygen. In fact, oxygen is Mother Nature's spin trap. To elucidate the role of free radicals in the oxygen uptake, we investigated the effect of the spin trap 5,5-dimethyl-1-pyrroline *N*-oxide (DMPO). Because DMPO is a radical trap, it is expected to be an antioxidant.

Studies with a Clark oxygen electrode showed that oxygen is consumed in incubations of cysteine and lactoperoxidase. The rate of oxygen consumption decreased with the addition of DMPO either during or before the reaction. These results are consistent with the cysteine thiyl free radical being formed and then trapped by DMPO before further reactions of the cysteine thiyl radical can lead to oxygen consumption. The reaction of radicals with DMPO is fast enough that the inhibition is complete. As was found with ESR, the addition of catalase also inhibited oxygen upake (Mottley et al. 1987).

The generation of hydrogen peroxide and the consumption of oxygen during the lactoperoxidase-catalyzed oxidation of cysteine can be explained on the basis of known reactions of the cysteine thiyl radical:

$$R\dot{S} + RSH(RS^-) \leftrightharpoons (RS \therefore SR)^- + H^+$$

$$(RS \therefore SR)^- + O_2 \rightarrow RSSR + \dot{O}_2^-$$

$$2\,\dot{O}_2^- + 2H^+ \rightarrow H_2O_2 + O_2$$

The thiyl free radical reacts with thiol anion to form the L-cystine disulphide anion free radical, which is air oxidized to L-cystine, forming superoxide. Even in the absence of superoxide dismutase, superoxide disproportionates rapidly to form hydrogen peroxide, which will drive the HRP-catalyzed reaction that forms thiyl radicals and, ultimately, more superoxide, resulting in an enzymatic free-radical chain reaction. The peroxidase-catalyzed oxidation of cysteine may be of physiologic significance. It is particularly noteworthy that cysteine oxidation by lactoperoxidase is independent of exogenously added hydrogen peroxide because cellular hydrogen peroxide concentrations are generally thought to be low, less than 0.1 μM (Sies et al. 1973).

Carbon-centered Free Radicals

The analog of the perhydroxyl radical ($\dot{O}OH$) in which the H atom is replaced by an organic group is called a peroxyl radical ($RO\dot{O}$). It has been said that peroxyl radicals do not dissociate to $\dot{O}_2^-$ because of the stability of the carbon-oxygen bond (Marnett 1987). Although this statement is generally true, exceptions do exist. The decomposition of a peroxyl free radical to superoxide normally would require the formation of a thermodynamically unfavorable cabonium ion.

$$-\overset{|}{\underset{|}{C}}-O\dot{O} \rightarrow -\overset{|}{\underset{|}{C^+}} + \dot{O}_2^-$$

If a more stable carbon species can be formed, then superoxide formation can occur from the decomposition of peroxyl radicals.

The oxidation of indole-3-acetic acid by HRP forms the skatole carbon-centered free radical (Mottley and Mason 1986). In the presence of oxygen, superoxide is also detected. The skatole carbon-centered free radical is oxidized via the reaction shown below, producing the superoxide radical.

$CH_2\bullet$ (indole, NH) $+ O_2 \rightarrow$ $CH_2OO\bullet$ (indole, NH) $\rightarrow$ CH_2 (3-methyleneindolenine, N) $+ O_2\bullet^- + H^+$

This reaction is similar to that proposed for the formation of superoxide radical from the α-hydroxyethyl radical (Reszka and Chignell 1983) and the α-hydroxybenzyl radical (Hammel et al. 1985) and to the autoxidation of 1,4-cyclohexadiene to benzene (Howard and Ingold 1967). All these reactions occur within the formation of superoxide radical.

Semiquinones

In general, the semiquinones formed from stable hydroquinones do not reduce molecular oxygen to superoxide at a rapid rate, the equilibrium constant for this reaction being far to the left. Hydroquinone is the classic example of such a compound in that benzosemiquinone formation does not result in oxygen consumption (Ohnishi et al. 1969).

An interesting exception to this rule is diethylstilbestrol. Oxidation of diethylstilbestrol by the HRP system results in superoxide generation and in the superoxide-dependent production of DNA strand breaks (Epe et al. 1986). The superoxide formation is probably the result of the diethylstilbestrol semiquinone reacting with molecular oxygen to form the corresponding quinone and superoxide.

Free Radical Metabolites That Form Superoxide by Oxidizing Glutathione

As discussed previously, the chemistry of thiyl free radicals leads to the formation of superoxide free radicals and, under some conditions, of per-

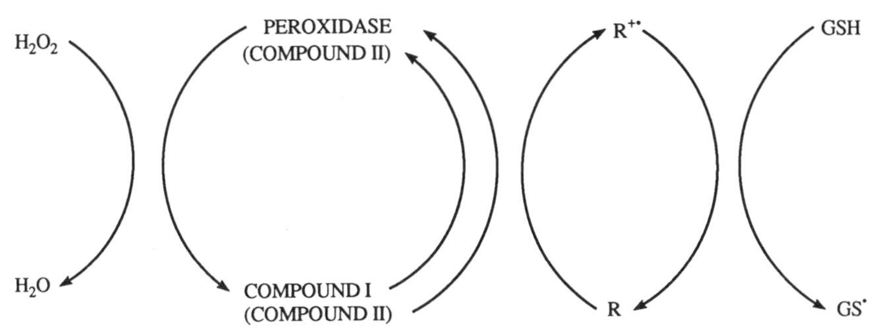

Figure 2-7 Futile metabolism of a wide variety of compounds that are oxidized by peroxidases to free radical metabolites, which are then reduced to the parent compounds by glutathione (GSH).

oxyl free radicals. Many free radical metabolites react with thiols, including glutathione (GSH), to form thiyl free radicals (Figure 2-7) while regenerating the parent compound that had been metabolized to a free radical. This process is a futile metabolism where the free radical precusor is a catalyst of thiyl radical formation. The first report of this chemical process, sometimes called thiol pumping, was from the laboratory of Yamazaki and coworkers, who showed that GSH reacts with the chlorpromazine radical cation to produce the GṠ radical (Ohnishi et al. 1969). The literature on thiol pumping has been reviewed by Ross (1988).

The GSH thiyl radical is also formed in an aminopyrine-catalyzed reaction (Eling et al. 1985). Aminopyrine is oxidized by prostaglandin H synthase hydroperoxidase to the aminopyrine radical cation. In the presence of GSH, the aminopyrine radical cation is reduced, resulting in the formation of aminopyrine and the GṠ radical. The phenoxyl radical of diethylstilbestrol, which is proposed as a possible determinant of genotoxicity of this compound, has also been shown to react with GSH to generate the GṠ radical (Ross et al. 1985b).

Ross and coworkers have done extensive studies on the fate of GSH in HRP and prostaglandin H synthase-catalyzed oxidations of *p*-phenetidine and acetaminophen (Ross et al. 1985b). They have shown that the free radical metabolites of both *p*-phenetidine and acetaminophen react readily with GSH to form the parent molecules and the GṠ radical, the latter being trapped by DMPO.

Recent pulse-radiolysis experiments failed to detect any reaction between the acetaminophen phenoxyl radical and cysteine (Bisby and Tabassum 1988).

$$\dot{O}C_6H_5NHAC + RSH \leftrightharpoons R\dot{S} + HOC_6H_5NHAC$$

The reduction potential of a thiol group is +0.83 V (61), which is about 0.1 V higher than that for the acetaminophen phenoxyl radical; hence, the reverse reaction, the reduction of the thiyl radical by acetaminophen, is thermodynamically favored. Wilson and coworkers (1986) have studied a similar phenomenon involving the reaction between the aminopyrine radical cation and GSH: $\dot{AP}^+ + GSH \leftrightharpoons AP + G\dot{S}$. They have shown that, thermodynamically, the equilibrium is on the left-hand side of the reaction. This reaction apparently proceeds to the right only because the thiyl radical reacts with thiolate or oxygen or both, ultimately forming oxidized glutathione (GSSG).

We decided to elucidate this reaction using direct ESR spectroscopy with the fast-flow technique (Rao et al. 1990). This approach is more direct than the spin-trapping experiment. On theoretical grounds, the thiyl free radical cannot be detected directly with ESR. Nevertheless, the existence of thiyl radicals can be demonstrated because thiyl radicals react with glutathione thiol anion to form a detectable glutathione disulfide anion radical (Figure 2-8). Because this free radial is air-oxidized to glutathione disulfide, forming superoxide (Figure 2-9), experiments must be done under nitrogen. The ESR literature further indicated that $RSO\dot{O}$ radicals (Figure 2-9, equation 5) can also form as a result of an addition reaction of thiyl radicals with oxygen (Sevilla et al. 1988, 1989). Indeed, we could demonstrate glutathione disulfide anion radical formation $(GS \therefore SG)^-$ (see Figure 2-8, upper) by flowing acetaminophen and H_2O_2 against GSH and HRP (Rao et al. 1990). We observed two ESR spectra with different g-values. The triplet with some hyperfine structure at $g = 2.0043$ is due to the acetaminophen phenoxyl free radical (Fischer et al. 1986), whereas the quintet, with its fourth and fifth lines hidden under the triplet, has a g-value of 2.013, which is typical for the disulfide anion radical. The quintet results from a coupling of the free electron with four equivalent hydrogens. These are the four methylene hydrogens next to the sulfur atoms. A control experiment without acetaminophen demonstrated that the glutathione disulfide anion radical formation is not due to direct oxidation of GSH by the enzymatic system. Omission of hydrogen peroxide or HRP prevents any radical formation, confirming the dependence on the enzymatic system. Omission of GSH yields an increase in the acetaminophen free radical signal, showing that GSH does reduce the acetaminophen phenoxyl free radical. In the presence of oxygen, oxidized glutathione comes primarily from its disulfide anion radical and not from the dimerization of thiyl radicals (see Figure 2-9, equation 3), as is often stated.

Ascorbate is a better reducing agent than GSH (see Figure 2-9, equation 2). It totally inhibits the acetaminophen free radical formation in the same enzymatic system we used for the GSH or cysteine incubations. Omission of acetaminophen shows that ascorbate is also a substrate for the enzymatic system, but the rate of oxidation is much slower (Rao et al. 1990). Addition of both GSH and ascorbate leads only to the ascorbate free radical, implying ascorbate, and not GSH, will be oxidized by the acetaminophen free radical in those cells that contain both reducing agents (Rao et al. 1990).

The HRP-catalyzed oxidation of certain drugs in the presence of GSH results in extensive oxygen consumption (Subrahmanyam and O'Brien 1985;

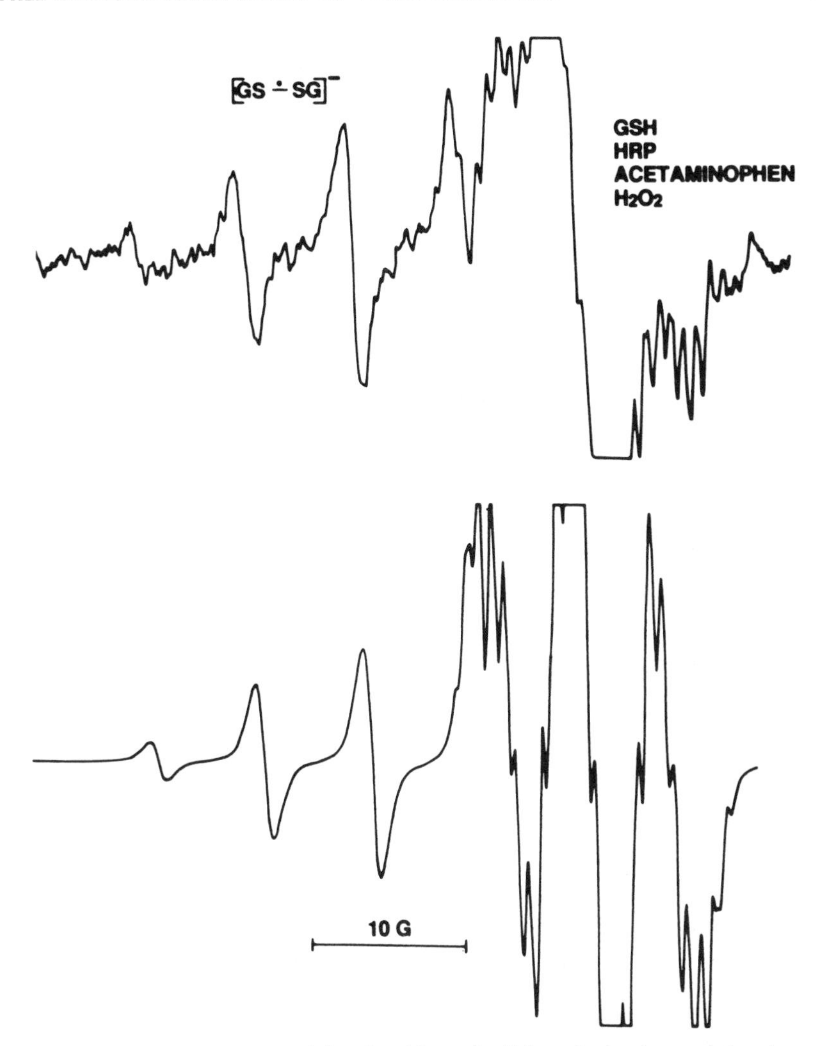

Figure 2-8 The ESR spectrum of the glutathione disulfide radical anion and the phenoxyl radical of acetaminophen generated in a system of glutathione (100 mM), horseradish peroxidase type I (10 units/ml), hydrogen peroxide (25 mM), and acetaminophen (10 mM) in phosphate (0.1 M) buffer, pH 7.5. Below is the computer simulation of the glutathione disulfide radical anion and the acetaminophen phenoxyl free radical (Rao et al. 1990).

O'Brien 1988). Only catalytic amounts of drugs were required for this futile metabolism mechanism (see Figure 2-9, equation 1). Active drugs included phenothiazine, aminopyrine, *p*-phenetidine, acetaminophen, and 4-*N*,*N*-$(CH_3)_2$-aminophenol (Subrahmanyam et al. 1987). A stoichiometry of 0.4 to 0.5 mol of O_2 was consumed per mole of GSH. Oxygen uptake was also stimulated up to twofold by superoxide dismutase. Other drugs, including dopamine and α-methyl dopa, did not catalyze oxygen uptake, nor were GSSG or thiyl radicals formed. Instead, GSH was depleted by GSH conjugate formation (Subrahmanyam et al. 1987).

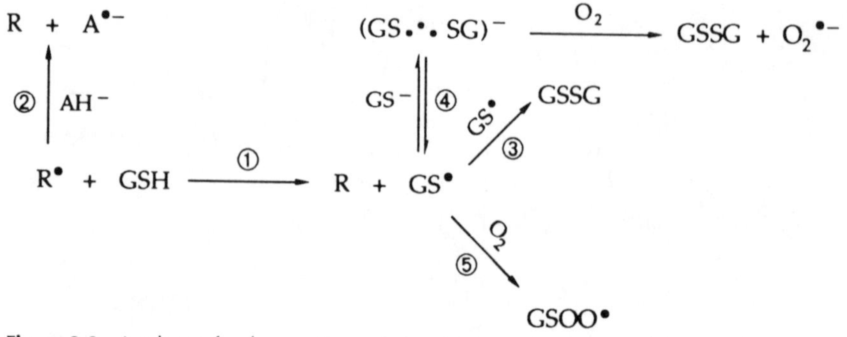

Figure 2-9 A scheme for the reactions of glutathione and ascorbate with many free radical metabolites formed by peroxidases and the subsequent reactions of the glutathione thiyl free radical.

The peroxidase concentration could markedly affect GSH-mediated oxygen activation. Thus, oxygen uptake occurred only with low peroxidase concentrations in the case of *p*-phenetidine and acetaminophen. When GSH oxidation products were analyzed by HPLC, it was found that a stoichiometric oxidation of GSH to GSSG accompanied the oxygen uptake. No GSH oxidation occurred when the reaction was carried out under argon (Subrahmanyam et al. 1987).

Free Radical Metabolites That Form Superoxide by Oxidizing NAD(P)H

NAD(P)H oxidation catalyzed by peroxidase in the presence of catalytic amounts of various xenobiotics and H_2O_2 is also accompanied by extensive oxygen uptake. No oxidation occurs in the absence of the xenobiotic. A stoichiometry of 0.8 to 1 mole of oxygen is consumed per mole of NADH oxidized (Subrahmanyam and O'Brien 1985; O'Brien 1988). NAD(P)H oxidation by catalytic amounts of benzidine, *o*-toluidine, *p*-aminophenol, or *p*-phenetidine with low peroxidase concentrations and catalytic amounts of H_2O_2 was accompanied by extensive oxygen uptake (O'Brien 1988). The free radical metabolites oxidize NAD(P)H to NAD(Ṗ) (Figure 2-10),

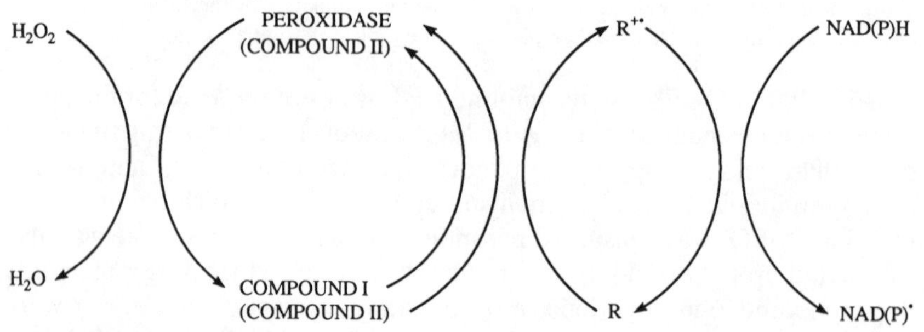

Figure 2-10 Futile metabolism of a wide variety of compounds that are oxidized by peroxidases to free radical metabolites, which are then reduced to the parent compound by NADPH.

which reacts with molecular oxygen at a nearly diffusion-limited rate (Land and Swallow 1971).

$$NAD(\dot{P}) + O_2 \rightarrow NAD(P)^+ + \dot{O}_2^-$$

Superoxide formation also increases in the presence of NADH in a diethylstilbestrol/HRP/H_2O_2 system (Epe et al. 1986).

XENOBIOTIC SOURCES OF PEROXYL FREE RADICALS

Trichloromethyl Radical and Other Carbon-centered Free Radical Metabolites

Perhaps the best known free radical metabolite of a xenobiotic substance is the trichloromethyl radical ($\dot{C}Cl_3$), first proposed by Butler (1961) and confirmed by Poyer and colleagues (1980) using spin trapping and [^{13}C]-carbon tetrachloride. Using pulse radiolysis, the $\dot{C}Cl_3$ radical has been shown to react rapidly with oxygen to form the $Cl_3CO\dot{O}$ peroxyl radical (Packer et al. 1978). Because carbon tetrachloride is metabolized to phosgene and carbon dioxide, it is reasonable to assume that the $Cl_3CO\dot{O}$ radical is formed in biologic systems, but direct ESR evidence is lacking.

Prostaglandin H synthase hydroperoxidase-mediated metabolism of the prostraglandin H synthase inhibitor phenylbutazone forms a phenylbutazone-derived, carbon-centered radical (Hughes et al. 1988). The 4-hydroperoxy metabolite was isolated from incubations of phenylbutazone with either ram seminal vesicle microsomes or HRP. The phenylbutazone peroxyl radical is proposed to be the actual inhibitor of prostaglandin H synthase. The 4-hydroperoxyphenylbutazone was not reduced to 4-hydroxyphenylbutazone by prostaglandin H hydroperoxidase, but it is proposed that the 4-hydroxyphenylbutazone formation occurs by a nonenzymatic reaction of two phenylbutazone peroxyl radicals and their subsequent rearrangement to alkoxy radicals, which abstract hydrogen atoms (Hughes et al. 1988). Hydroxylation of substituted hydrazines by cytochrome P-450 to form an *N*-hydroxy intermediate followed by the loss of water forms a monosubstituted diazine:

$$R{-}NH{-}NH_2 \xrightarrow{\text{P-450}} R{-}NH{-}N(OH)H \xrightarrow{-H_2O} R{-}N{=}N{-}H$$

If air oxidation and decomposition of these diazines proceed as has been proposed for phenylhydrazine, then carbon-centered free radicals and superoxide will be formed.

$$R{-}N{=}N{-}H + O_2 \rightarrow \dot{R} + N_2 + \dot{O}_2^- + H^+$$

Procarbazine is oxidized in a similar way by HRP to form both the methyl and benzylic-type $RC_6H_4\text{-}\dot{C}H_2$ carbon-centered radicals (Sinha 1984). An oxygen-centered adduct with DMPO was also formed in this system and was assigned to a peroxyl radical formed from the reaction of oxygen with one of the carbon-centered radicals. Iproniazid can also be oxidized to free radicals by either HRP or prostaglandin hydroperoxidase (Sinha 1983). The oxidation of iproniazid by either peroxidase formed a carbon-centered free radical identified as the isopropyl radical, which reacts with oxygen to form an isopropyl peroxyl radical (Sinha 1983).

Sulfite-Derived, Oxygen-Centered Free Radicals

The one-electron oxidation of (bi)sulfite is catalyzed by peroxidase:

$$SO_3^{2-} \rightarrow \dot{S}O_3^- + e$$

Either HRP (Araiso et al. 1976; Mottley et al. 1982b) or the prostaglandin H synthase (hydroperoxidase) in ram seminal vesicles or guinea pig lung microsomes (Mottley et al. 1982a) catalyzes the formation of the sulfur trioxide anion free radical ($\dot{S}O_3^-$), as demonstrated with ESR investigations. The initiating one-electron oxidation yields the sulfur trioxide radical anion ($\dot{S}O_3^-$), a predominantly sulfur-centered radical that reacts with molecular oxygen. Superoxide initiates (bi)sulfite oxidation by forming $\dot{S}O_3^-$ (McCord and Fridovich 1968, 1969) but is not a chain propagator (McCord and Fridovich 1968; Hayon et al. 1972; Huie and Neta 1985), implying that superoxide is not a significant reaction product of the $\dot{S}O_3^-$ radical. The major reaction product of $\dot{S}O_3^-$ with oxygen is proposed to be a peroxyl free radical (Hayon et al. 1972; Reed et al. 1986): $\dot{S}O_3^- + O_2 \rightarrow {}^-O_3SO\dot{O}$. We found that this peroxyl free radical reacts with (±)-*trans*-7,8-dihydroxy-7,8-dihydrobenzo[*a*]pyrene to form diolepoxides, ultimate carcinogenic derivatives of benzo [*a*] pyrene (Reed et al. 1986).

The next step in the reaction sequence is the formation of the oxygen-centered sulfate anion radical from the peroxyl radical. Definitive proof of the existence of this radical in sulfite oxidation came through spin trapping with DMPO (Mottley and Mason 1988). In this work the initiation step was an enzymatic oxidation by HRP and hydrogen peroxide rather than autoxidation, but the rest of the mechanism is the same. The sulfur trioxide anion radical was trapped initially at high concentrations of DMPO (270 mM). That this radical was sulfur-centered was proved using ^{33}S-labeled sulfite, resulting in the ^{33}S hyperfine structure observed in the ESR spectrum (Mottley and Mason 1988). To trap possible secondary radicals, $\dot{O}OSO_3^-$ and $S\dot{O}_4^-$, the concentration of DMPO was decreased so that not all the $\dot{S}O_3^-$ radical would be trapped, but would instead react further. Using a DMPO concentration of 3 mM, the ESR spectrum of the DMPO/$S\dot{O}_4^-$ radical adduct was detected.

The sulfate anion radical is known to react with many compounds more commonly thought of as hydroxyl radical scavengers such as formate and ethanol. Free radicals derived from these scavengers are trapped in systems where sulfite peroxidation has been inhibited by these scavengers (Mottley and Mason 1988).

FREE RADICAL PROMOTION FROM ORGANIC HYDROPEROXIDES BY HEMOPROTEINS

Both hematin and hemoproteins react with organic hydroperoxides, making the role of the apoprotein in these reactions unclear. Three different mechanisms, each predicting both peroxyl and alkoxyl radical formation, have been proposed. In the first mechanism, hematin catalyzes the decomposition of hydroperoxides through the same redox reactions reported for transition metal ions:

$$\text{Hematin}(Fe^{3+}) + ROOH \rightarrow \text{hematin}(Fe^{2+}) + RO\dot{O} + H^+$$

$$\text{Hematin}(Fe^{2+}) + ROOH \rightarrow \text{hematin}(Fe^{3+}) + R\dot{O} + OH^-$$

The second reaction is essentially the classical Fenton reaction with hydrogen peroxide replaced by the organic hydroperoxide. Tappel (1961) proposed an alternative scheme whereby the hydroperoxides are broken down by hematin or cytochrome c in a reaction in which no change in the valency of iron occurs, but which involves instead the homolytic scission of the oxygen-oxygen bond to form an iron-bound hydroxyl radical (Cadenas et al. 1980):

$$\text{Hematin}(Fe^{3+}) + ROOH \rightarrow \text{hematin}(Fe^{3+})-\dot{O}H + R\dot{O}$$

$$\text{Hematin}(Fe^{3+})-\dot{O}H + ROOH \rightarrow \text{hematin}(Fe^{3+}) + RO\dot{O} + H_2O$$

Overall, the net result of these two mechanisms is the same:

$$2ROOH \rightarrow R\dot{O} + RO\dot{O} + H_2O$$

The third mechanism is suggested by the report of the oxidation of deuteroferriheme by hydrogen peroxide or peroxobenzoic acids to a species analogous to compound I of peroxidases and catalase (Jones et al. 1977). If this species oxidizes organic hydroperoxides, as catalase compound I is thought to do (Chance 1952), then the peroxyl radical will be formed:

$$\text{Catalase compound I} + ROOH \rightarrow RO\dot{O} + \text{catalase compound II}$$

The *tert*-butylperoxyl and cumylperoxyl radicals have been observed directly by ESR in a hematin-hydroperoxide system (Kalyanaraman et al. 1983) and characterized by their distinctive g-values ($g = 2.014$). Spin trapping evidence for peroxyl free radical formation is also convincing (Mottley and Mason 1989). One area of current interest is the study of the reaction of organic hydroperoxides in red blood cells. The administration of peroxides to red blood cells in vitro has been shown to cause lipid peroxidation in cell membranes, increased passive cation permeability, depletion of glutathione, decreased cellular and membrane deformability, increased membrane-associated protein crosslinking and osmotic fragility, the oxidative denaturation of hemoglobin, an increase in flux through the hexose monophosphate shunt, and free radical formation (Maples et al. 1990 and references therein).

DMPO has been used to trap *t*-butyloxy radicals formed in oxyhemoglobin-containing suspensions of erythrocytes in the presence of *t*-butylhydroperoxide (Thornalley et al. 1983). Free radical production in intact cell systems was slower than that in hemoglobin solutions, indicating that membrane transport of *t*-butyl hydroperoxide may be a rate-determining step. Although the *t*-butyloxy radicals are rapidly scavenged by DMPO, 100 mM DMPO decreased the rate of oxyhemoglobin oxidation by only about 30%. This failure of a 1000-fold excess of DMPO over oxyhemoglobin and *t*-butyl hydroperoxide to stop the hemoglobin oxidation suggests that the *t*-butyloxy radical may not be the major oxidant of oxyhemoglobin. Thornally and coworkers (1983) suggest that the oxyhemoglobin may be oxidized in the free radical forming process:

$$Hb(Fe^{II}\text{-}O_2) + t\text{-BuOOH} \rightarrow Hb(Fe^{III}) + O_2 + t\text{-Bu}\dot{O} + OH^-$$

Cumene and *t*-butyl hydroperoxide are tumor promoters in skin. This activity is presumed to be mediated through the activation of these hydroperoxides to free radicals (Marnett 1987). DMPO/t-Bu$\dot{O}$ and DMPO/$\dot{C}H_3$ were detected shortly after the addition of *t*-butylhydroperoxide to the target cell, the murine basal keratinocyte (Taffe et al. 1987). The radical-generating capacity of subcellular fractions of these epidermal cells was limited to the 105,000 × g supernatant fraction and was heat labile (Taffe et al. 1987). In both erythrocytes and keratinocytes, reduction of hydroperoxides to alkoxyl free radicals appears to be the most probable initial radical forming reaction.

In contrast, chloroperoxidase catalyzes the oxidation of organic hydroperoxides (ROOH) to their peroxyl radicals (Chamulitrat et al. 1989).

$$ROOH + \text{chloroperoxidase} \rightarrow ROH + \text{compound I}$$

$$\text{Compound I} + ROOH \rightarrow \text{compound II} + RO\dot{O}$$

$$\text{Compound II} + ROOH \rightarrow \text{chloroperoxidase} + RO\dot{O}$$

In the case of ethyl hydroperoxide, EtO$\dot{O}$ appears to react through an enzyme-independent bimolecular reaction to form a tetraoxygen intermediate originally proposed by Russell (Ingold 1969), which decomposes to produce acetaldehyde, ethanol, and molecular oxygen.

$$2\ EtO\dot{O} \rightarrow O_2 + CH_3CH_2OH + CH_3CHO$$

We have found that about 12% of the molecular oxygen formed occurs as excited singlet oxygen, $^1O_2$ (Hall et al. 1989).

CONCLUSION

Oxygen reacts with many free radical metabolites of toxic chemicals and drugs. In particular, radical anion metabolites react by electron transfer to form superoxide and regenerate the parent compound, resulting in an oxygen-dependent futile cycle. Other mainly neutral free radicals such as the trichloromethyl free radical react with oxygen by covalent bond formation to form peroxyl radicals. Radical cation metabolites and other free radicals formed by oxidation do not react rapidly with oxygen but oxidize GSH instead, forming a thiyl free radical and regenerating the parent compound in a GSH-dependent futile cycle. The chemistry of thiyl free radicals can result in both superoxide and peroxyl free radical production, although superoxide formation dominates in most circumstances.

The formation of oxygen-derived free radicals is clearly the root cause of the toxicity of many chemicals including paraquat and carbon tetrachloride, the classic pulmonary and hepatic toxins, respectively. Evidence that the free radical metabolites play a role in the side effects of drugs such as nitrofurantoin and anticancer quinones is also substantial. On the other hand, the role of organic peroxyl and alkoxyl free radicals in tumor promotion by organic hydroperoxides is not yet understood. In any case, free radical metabolites of toxic chemicals and drugs appear to be the most universal of the reactive intermediates ultimately responsible for damage to biochemical macromolecules. In every case, the reactions of these free radical metabolites ultimately lead to the formation of superoxide or peroxyl and alkoxyl radical formation. In many cases, it is these oxygen-centered radicals that actually appear to be responsible for the toxicity.

REFERENCES

Araiso, T., Miyoshi, K., and Yamazaki, I. (1976). Mechanisms of electron transfer from sulfite to horseradish peroxidase-hydroperoxide compounds. *Biochemistry*, 15: 3059–63.

Bisby, R. H., and Tabassum, N. (1988). Properties of the radicals formed by one-electron oxidation of acetaminophen—a pulse radiolysis study. *Biochem. Pharmacol.*, 37: 2731–38.

Boyd, M. R., Catignani, G.L., Sasame, H.A., Mitchell, J.R., and Stiko, A.W. (1979). Acute pulmonary injury in rats by nitrofurantion and modification by vitamin E, dietary fat, and oxygen. *Am. Rev. Respir. Dis.* 120: 93–99.

Butler, T. C. (1961). Reduction of carbon tetrachloride *in vivo* and reduction of carbon tetrachloride and chloroform *in vitro* by tissues and tissue constituents. *J. Pharmacol. Exp. Ther.*, 134: 311–19.

Cadenas, E., Boveris, A., and Chance, B. (1980). Low-level chemiluminescence of hydroperoxide-supplemented cytochrome *c*. *Biochem. J.*, 187: 131–40.

Chamulitrat, W., Takahashi, N., and Mason, R. P. (1989). Peroxyl, alkoxyl, and carbon-centered radical formation from organic hydroperoxides by chloroperoxidase. *J. Biol. Chem.*, 264: 7889–99.

Chance, B. (1952). The spectra of the enzyme-substrate complexes of catalase and peroxidase. *Arch. Biochem. Biophys.*, 41: 404–15.

DeGray, J. A., Rao, D. N. R., and Mason, R. P. (1991). Reduction of paraquat and related bipyridylium compounds to free radical metabolites by rat hepatocytes. *Arch. Biochem. Biophys.*, 289: 145–52.

Docampo, R., Moreno, S. N. J., and Mason, R. P. (1983). Generation of free radical metabolites and superoxide anion by the calcium indicators arsenazo III, antipyrylazo III, and murexide in rat liver microsomes. *J. Biol. Chem.*, 258: 14920–25.

Dodge, A. D. (1971). The mode of action of the bipyridylium herbicides, paraquat and diquat. *Endeavour*, 30: 130–35.

Eling, T. E., Mason, R. P., and Sivarajah, K. (1985). The formation of aminopyrine cation radical by the peroxidase activity of prostaglandin H synthase and subsequent reactions of the radical. *J. Biol. Chem.*, 260: 1601–07.

Epe, B., Schiffmann, D., and Metzler, M. (1986). Possible role of oxygen radicals in cell transformation by diethylstilbestrol and related compounds. *Carcinogenesis*, 7: 1329–34.

Fischer, V., Harrelson, W. J., Jr., Chignell, C. F., and Mason, R. P. (1984). Spectroscopic studies of cutaneous photosensitizing agents. V. Spin trapping and direct electron spin resonance investigations of the photoreduction of gentian (crystal) violet. *Photochem. Photobiophys.*, 7: 111–19.

Fischer, V., Harman, L. S., West, P. R., and Mason, R. P. (1986). Direct electron spin resonance detection of free radical intermediates during the peroxidase catalyzed oxidation of phenacetin metabolites. *Chem.-Biol. Interact.*, 60: 115–27.

French, J. K., Winterbourn, C. C., and Carrell, R. W. (1978). Mechanism of oxyhaemoglobin breakdown on reaction with acetylphenylhydrazine. *Biochem. J.*, 173: 19–26.

Gage, J. C. (1968). The action of paraquat and diquat on the respiration of liver cell fractions. *Biochem J.*, 109: 757–61.

Gant, T. W., Rao, D. N. R., Mason, R. P., and Cohen, G. M. (1988). Redox cycling and sulphydryl arylation; Their relative importance in the mechanism of quinone cytotoxicity to isolated hepatocytes. *Chem.-Biol. Interact.*, 65: 157–73.

Goldberg, B., Stern, A., and Peisach, J. (1976). The mechanism of superoxide anion generation by the interaction of phenylhydrazine with hemoglobin. *J. Biol. Chem.*, 251: 3045–51.

Goldberg, B., and Stern, A. (1977). The mechanism of oxidative hemolysis produced by phenylhdrazine. *Mol. Pharmacol.*, 13: 832–39.

Goldberg, B., Stern, A., Peisach, J., and Blumberg, W. E. (1979). The detection of superoxide anion from the reaction of oxyhemoglobin and phenylhydrazine using EPR spectroscopy. *Experientia*, 35: 488–89.

Hall, R. D., Chamulitrat, W., Takahashi, N., Chignell, C. F., and Mason, R. P. (1989). Detection of singlet ($^1O_2$) oxygen phosphorescence during chloroperoxidase-catalyzed decomposition of ethyl hydroperoxide. *J. Biol. Chem.*, 264: 7900–06.

Hammel, K. E., Tien, M., Kalyanaraman, B., and Kirk, T. K. (1985). Mechanism of oxidative C_α-C_B cleavage of a lignin model dimer by *Phanerochaete chrysosporium* ligninase. Stoichiometry and involvement of free radicals. *J. Biol. Chem.*, 260: 8348–53.

Handa, K., and Sato, S. (1975). Generation of free radicals of quinone group-containing anticancer chemicals in NADPH-microsome system as evidenced by initiation of sulfite oxidation. *Gann*, 66: 43–47.

Handa, K., and Sato, S. (1976). Stimulation of microsomal NADPH oxidation by quinone group-containing anticancer chemicals. *Gann*, 67: 523–28.

Harbour, J. R., and Bolton, J. R. (1975). Superoxide formation in spinach chloroplasts: Electron spin resonance detection by spin trapping. *Biochem. Biophys. Res. Commun.*, 64: 803–07.

Harrelson, W. G., Jr., and Mason, R. P. (1982). Microsomal reduction of gentian violet. Evidence for cytochrome P-450-catalyzed free radical formation. *Mol. Pharmacol.* 22: 239–42.

Hayon, E., Treinin, A., and Wilf, J. (1972). Electronic spectra, photochemistry, and autoxidation mechanism of the sulfite-bisulfite-pyrosulfite systems. The SO_2^-, SO_3^-, SO_4^-, and SO_5^- radicals. *J. Am. Chem. Soc.*, 94: 47–57.

Homer, R. F., Mees, G. C., and Tomlinson, T. E. (1960). Mode of action dipyridyl quaternary salts as herbicides. *J. Sci. Food Agric.*, 11: 309–15.

Horton, J. K., Brigelius, R., Mason, R. P., and Bend, J. R. (1986). Paraquat uptake into freshly isolated rabbit lung epithelial cells and its reduction to the paraquat radical under anaerobic conditions. *Mol. Pharmacol.*, 29: 484–88.

Howard, J. A., and Ingold, K. U. (1967). Absolute rate constants for hydrocarbon autoxidation. V. The hydroperoxy radical in chain propagation and termination. *Canadian Journal of Chemistry*, 45: 785–92.

Hughes, M. F., Mason, R. P., and Eling, T. E. (1988). Prostaglandin hydroperoxidase-dependent oxidation of phenylbutazone: Relationship to inhibition of prostaglandin cyclooxygenase. *Mol. Phamacol.*, 34: 186–93.

Huie, R. E., and Neta, P. (1985). Oxidation of ascorbate and a tocopherol analogue by the sulfite-derived radicals $\dot{S}O_3^-$ and $S\dot{O}_5^-$. *Chem.-Biol. Interact.*, 53: 233–38.

Ingold, K. U. (1969). Peroxy radicals. *Accounts of Chemical Research*, 2: 1–9.

Jain, S. K., and Subrahmanyam, D. (1978). On the mechanism of phenylhydrazine-induced hemolytic anemia. *Biochem. Biophys. Res. Commun.*, 82: 1320–24.

Jain, S. K., and Hochstein, P. (1979). Generation of superoxide radicals by hydrazine, its role in phenylhydrazine-induced hemolytic anemia. *Biochim. Biophys. Acta*, 586: 128–36.

Jones, P., Mantle, D., Davies, D. M., and Kelly, H. C. (1977). Hydroperoxidase activities of ferrihemes: Heme analogues of peroxidase enzyme intermediates. *Biochemistry*, 16: 3974–78.

Jülich, T., Scheffler, K., Schuler, P., and Stegmann, H. B. (1988). EPR and ENDOR investigation of daunomycin and adriamycin semiquinones. *Magnetic Resonance in Chemistry*, 26: 701–06.

Kalyanaraman, B., Mottley, C., and Mason, R. P. (1983). A direct electron spin resonance and spin-trapping investigation of peroxyl free radical formation by hematin/hydroperoxide systems, *J. Biol. Chem.*, 258: 3855–58.

Kappus, H., and Sies, H. (1981). Toxic drug effects associated with oxygen metabolism. Redox cycling and lipid peroxidation. *Experientia*, 37: 1233–41.

Land, E. J., and Swallow, A. J. (1971). One-electron reactions in biochemical systems as studied by pulse radiolysis IV. Oxidation of dihydronicotinamide-adenine dinucleotide. *Biochim. Biophys. Acta*, 234: 34–42.

Liehr, J. G., and Roy, D. (1990). Free radical generation by redox cycling of estrogens. *Free Radic. Biol. Med.*, 8: 415–23.

Maples, K. R., Kennedy, C. H., Jordan, S. J., and Mason, R. P. (1990). *In vivo* thiyl free radical formation from hemoglobin following administration of hydroperoxides. *Arch. Biochem. Biophys.*, 277: 402–09.

Marnett, L. J. (1987). Peroxyl free radicals: potential mediators of tumor initiation and promotion. *Carcinogenesis*, 8: 1365–73.

Mason, R. P., and Holtzman, J. L. (1975a). The mechanism of microsomal and mitochondrial nitroreductase. Electron spin resonance evidence for nitroaromatic free radical intermediates. *Biochemistry*, 14: 1626–32.

Mason, R. P., and Holtzman, J. L. (1975b). The role of catalytic superoxide formation in the O_2 inhibition of nitroreductase. *Biochem. Biophys.Res. Commun.*, 67: 1267–74.

Mason, R. P., Peterson, F. J., and Holtzman, J. L. (1977). The formation of an azo anion free radical metabolite during the microsomal azo reduction of sulfonazo III. *Biochem. Biophys. Res. Commun.*, 75: 532–40.

Mason, R. P., Peterson, F. J., and Holtzman, J. L. (1978). Inhibition of azoreductase by oxygen. The role of the azo anion free radical metabolite in the reduction of oxygen to superoxide. *Mol. Pharmacol.*, 14: 665–71.

Mason, R. P. (1979). Free radical metabolites of foreign compounds and their toxicological significance. In: E. Hodgson, J. R. Bend, R. M. Philpot, eds., *Review in Biochemical Toxicology*, vol. 1. New York: Elsevier, pp. 151–200.

Mason, R. P., and Chignell, C. F. (1981). Free radicals in pharmacology and toxicology—selected topics. *Pharmacol. Rev.*, 33: 189–211.

Mason, R. P. (1982). Free-radical intermediates in the metabolism of toxic chemicals. In W. A. Pryor, ed. *Free Radicals in Biology*, vol. V. New York: Academic Press, pp. 161–222.

McCord, J. M., and Fridovich, I. (1968). The reduction of cytochrome *c* by milk xanthine oxidase. *J. Biol. Chem.*, 243: 5753–60.

McCord, J. M., and Fridovich, I. (1969). The utility of superoxide dismutase in studying free radical reactions. I. Radicals generated by the interaction of sulfite, dimethyl sulfoxide and oxygen. *J. Biol. Chem.*, 244: 6056–63.

Michaelis, L., and Hill, E. S. (1933). Potentiometric studies on semiquinones. *J. Am. Chem. Soc.*, 55: 1481–94.

Misra, H. P., and Fridovich, I. (1976). The oxidation of phenylhydrazine: Superoxide and mechanism. *Biochemistry*, 15: 681–87.

Moreno, S. N. J., Docampo, R., Mason, R. P., Leon, W., and Stoppani, A. O. M. (1982). Different behaviors of benznidazole as free radical generator with mammalian and *Trypanosoma cruzi* microsomal preparations. *Arch. Biochem. Biophys.*, 218: 585–91.

Moreno, S. N. J., Mason, R. P., and Docampo, R. (1984). Ca^{2+} and Mg^{2+}-enhanced reduction of arsenazo III to its anion free radical metabolite and generation of superoxide anion by an outer mitochondrial membrane azoreductase. *J. Biol. Chem.*, 259: 14609–16.

Moreno, S. N. J., Mason, R. P., and Docampo, R. (1985). Reduction of the metallochromic indicators arsenazo III and antipyrylazo III to their free radical metabolites by cytoplasmic enzymes. *FEBS Lett.*, 180: 229–33.

Mottley, C., Mason, R. P., Chignell, C. F., Sivarajah, K., and Eling, T. E. (1982a). The formation of sulfur trioxide radical anion during the prostaglandin hydroperoxidase-catalyzed oxidation of bisulfite (hydrated sulfur dioxide). *J. Biol. Chem.*, 257: 5050–55.

Mottley, C., Trice, T. B., and Mason, R. P. (1982b). Direct detection of the sulfur trioxide radical anion during the horseradish peroxidase-hydrogen peroxide oxidation of sulfite (aqueous sulfur dioxide). *Mol. Pharmacol.*, 22: 732–37.

Mottley, C., and Mason, R. P. (1986). An electron spin resonance study of free radical intermediates in the oxidation of indole acetic acid by horseradish peroxidase. *J. Biol. Chem.*, 261: 16860–64.

Mottley, C., Toy, K., and Mason, R. P. (1987). Oxidation of thiol drugs and biochemicals by the lactoperoxidase/hydrogen peroxide system. *Mol. Pharmacol.*, 31: 417–21.

Mottley, C., and Mason, R. P. (1988). Sulfate anion free radical formation by the peroxidation of (bi)sulfite and its reaction with hydroxyl radical scavengers. *Arch. Biochem. Biophys.*, 267: 681–89.

Mottley, C., and Mason, R. P. (1989). Nitroxide radical adducts in biology: Chemistry, applications, and pitfalls. In L.J. Berliner, and J. Reuben, ed. *Biological Magnetic Resonance*. New York: vol. 8: Plenum, pp. 489–546.

O'Brien, P. J. (1988). Radical formation during the peroxidase catalyzed metabolism of carcinogens and xenobiotics: The reactivity of these radicals with GSH, DNA, and unsaturated lipid. *Free Radic. Biol Med.*, 4: 169–83.

Ohnishi, T., Yamazaki, H., Iyanagi, T., Nakamura, T., and Yamazaki, I. (1969). One-electron-transfer reactions in biochemical systems: II. The reaction of free radicals formed in the enzymic oxidation. *Biochim. Biophys. Acta*, 172: 357–69.

Olsen, J., and Davis L. (1976). The oxidation of dithiothreitol by peroxidases and oxygen. *Biochim. Biophys. Acta*, 445: 324–29.

Packer, J. E., Slater, T. F., and Willson, R. L. (1978). Reactions of the carbon tetrachloride-related peroxy free radical ($CC1_3\dot{O}_2$) with amino acids: Pulse radiolysis evidence. *Life Sci.* 23: 2617–20.

Peterson, F. J., Combs, G. F., Jr., Holtzman, J. L., and Mason, R. P. (1982). Effect of selenium and vitamin E deficiency on nitrofurantoin toxicity in the chick. *J. Nutr.*, 112: 1741–46.

Powis, G. (1989). Free radical formation by antitumor quinones. *Free Radic. Bio. Med.*, 6: 63–101.

Poyer, J. L., McCay, P. B., Lai, E. K., Janzen, E. G., and Davis, E. R. (1980). Confirmation of assignment of the trichloromethyl radical spin adduct detected by spin trapping during ^{13}C-carbon tetrachloride metabolism *in vitro* and *in vivo*. *Biochem. Biophys. Res. Commun.*, 94: 1154–60.

Rao, D. N. R., Harman, L., Motten, A., Schreiber, J., and Mason, R. P. (1987). Generation of radical anions of nitrofurantoin, misonidazole, and metronidazole by ascorbate. *Arch. Biochem. Biophys.*, 255: 419–27.

Rao, D. N. R., Jordan, S., and Mason, R. P. (1988a). Generation of nitro radical anions of some 5-nitrofurans, and 2- and 5-nitroimidazoles by rat heptocytes. *Biochem. Pharmacol.*, 37: 2907–13.

Rao, D. N. R., Takahashi, N., and Mason, R. P. (1988b). Characterization of a glutathione conjugate of the 1,4-benzosemiquinone-free radical formed in rat hepatocytes. *J. Biol. Chem.*, 263: 17981–86.

Rao, D. N. R., Fischer, V., and Mason, R. P. (1990). Glutathione and ascorbate reduction of the acetaminophen radical formed by peroxidase. Detection of the glutathione disulfide radical anion and the ascorbyl radical. *J. Biol. Chem.*, 265: 844–47.

Reed, G. A., Curtis, J. F., Mottley, C., Eling, T. E., and Mason, R. P. (1986). Epoxidation of (±)-7,8-dihydroxy-7,8-dihydrobenzo[*a*]pyrene during (bi)sulfite autoxidation: Activation of a procarcinogen by a cocarcinogen. *Proc. Natl. Acad. Sci. USA*, 83: 7499–502.

Reszka, K., and Chignell, C. F. (1983). Spectroscopic studies of cutaneous photosensitizing agents-IV. The photolysis of benoxaprofen, an anti-inflammatory drug with phototoxic properties. *Photochem. Photobiol.*, 38: 281–91.

Rose, M. S., and Smith, L. L. (1977). Tissue uptake of paraquat and diquat. *Gen. Pharmacol.* 8: 173–76.

Ross, D., Larsson, R., Andersson, B., Nilsson, U., Lindquist, T., Lindeke, B., and Moldéus, P. (1985a). The oxidation of *p*-phenetidine by horseradish peroxidase and prostaglandin synthase and the fate of glutathione during such oxidations. *Biochem. Pharmacol.*, 34: 343–51.

Ross, D., Mehlhorn, R. J., Moldeus, P., and Smith, M. T. (1985b). Metabolism of diethylstilbestrol by horseradish peroxidase and prostaglandin-H synthase. Generation of a free radical intermediate and its interaction with glutathione. *J. Biol. Chem.*, 260: 16210–14.

Ross, D. (1988). Glutathione, free radicals and chemotherapeutic agents. Mechanisms of free-radical induced toxicity and glutathione-dependent protection. *Pharmacol. Ther.*, 37: 231–49.

Schreiber, J., Mottley, C., Sinha, B. K., Kalyanaraman, B., and Mason, R. P. (1987). One-electron reduction of daunomycin, daunomycinone, and 7-deoxydaunomycinone by the xanthine/xanthine oxidase system: Detection of semiquinone free radicals by electron spin resonance. *J. Am. Chem. Soc.*, 109: 348–51.

Sevilla, M. D., Yan, M., and Becker, D. (1988). Thiol peroxyl radical formation from the reaction of cysteine thiyl radical with molecular oxygen: An ESR investigation. *Biochem. Biophys. Res. Commun.*, 155: 405–10.

Sevilla, M. D., Yan, M., Becker, D., and Gillich, S. (1989). ESR investigations of the reactions of radiation-produced thiyl and DNA peroxyl radicals: Formation of sulfoxyl radicals. *Free Radic. Res. Comms.*, 6: 99–102.

Sies, H., Bücher, T., Oshino, N., and Chance, B. (1973). Heme occupancy of catalase in hemoglobin-free perfused rat liver and of isolated rat liver catalase. *Arch. Biochem. Biophys.*, 154: 106–16.

Sinha, B. K. (1983). Enzymatic activation of hydrazine derivatives. A spin trapping study. *J. Biol. Chem.*, 258: 796–801.

Sinha, B. K. (1984). Metabolic activation of procarbazine: Evidence for carbon-centered free-radical intermediates. *Biochem. Pharmacol.*, 33: 2777–81.

Sinha, B. K. (1989). Free radicals in anticancer drug pharmacology. *Chem.-Biol. Interact.*, 69: 293–317.

Smith, P., and Maples, K. R. (1985). EPR study of the oxidation of phenylhydrazine initiated by the titanous chloride/hydrogen peroxide reaction and by oxyhemoglobin. *J. Magnetic Resonance*, 65: 491–96.

Subrahmanyam, V. V., and O'Brien, P. J. (1985). Peroxidase catalysed oxygen activation by arylamine carcinogens and phenol. *Chem.-Biol. Interact.*, 56: 185–99.

Subrahmanyam, V. V., McGirr, L. G., and O'Brien, P. J. (1987). Glutathione oxidation during peroxidase catalysed drug metabolism. *Chem.-Biol. Interact.*, 61: 45–59.

Surdhar, P. S., and Armstrong, D. A. (1986). Redox potentials of some sulfur-containing radicals. *Journal of Physical Chemistry*, 90: 5915–17.

Taffe, B. G., Takahashi, N., Kensler, T. W., and Mason, R. P. (1987). Generation of free radicals from organic hydroperoxide tumor promoters in isolated mouse keratinocytes; formation of alkyl and alkoxyl radicals from *tert*-butyl hydroperoxide and cumene hydroperoxide. *J. Biol. Chem.*, 262: 12143–49.

Tappel, A. L. (1961). Biocatalysts: Lipoxidase and hematin compounds. In W. O. Lundberg, ed. *Autoxidation and Autoxidants*, New York: Interscience, vol. 1, pp. 325–66.

Thornalley, P. J., Trotta, R. J., and Stern A. (1983). Free radical involvement in the oxidative phenomena induced by *tert*-butyl hydroperoxide in erythrocytes. *Biochim. Biophys. Acta*, 759: 16–22.

Valenzuela, A., Rios, H., and Neiman, G. (1977). Evidence that superoxide radicals are involved in the hemolytic mechanism of phenylhydrazine. *Experientia*, 33: 962–63.

Wilson, I., Wardman, P., Cohen, G. M., and Doherty, M. D. (1986). Reductive role of glutathione in the redox cycling of oxidizable drugs. *Biochem. Pharmacol.*, 35: 21–22.

3

Cellular Dysfunction Induced by Ischemia/Reperfusion: Role of Reactive Oxygen Metabolites and Granulocytes

RONALD J. KORTHUIS, DONNA L. CARDEN,
AND D. NEIL GRANGER

In recent years, much research has been directed at elucidating the mechanisms underlying the pathophysiologic alterations associated with reperfusion (reoxygenation) of ischemic (hypoxic) tissues. As a consequence of this intensive effort, a large body of evidence has accumulated implicating a role for reactive oxygen metabolites and activated granulocytes in the genesis of postischemic cellular dysfunction. Figure 3-1 summarizes a hypothesis that has been proposed to explain the interaction of xanthine oxidase-derived oxidants, granulocyte infiltration, and the microvascular and parenchymal cell dysfunction that occurs in postischemic tissues (Granger 1988; Carden and Korthuis 1989). According to this scheme, xanthine oxidase-derived oxidants, produced at reperfusion, initiate the formation and release of proinflammatory agents, which subsequently attract and activate granulocytes. The activated neutrophils adhere to the microvascular endothelium, extravasate, and release cytotoxic oxidants and proteases that contribute to tissue dysfunction. This concept was first described for the postischemic intestine and has subsequently been extended to a number of other organ systems including the heart, skeletal muscle, stomach, pancreas, liver, kidney, brain, and skin. This review summarizes the supportive evidence and identifies areas of controversy and uncertainty regarding each component of the scheme.

THE CONCEPT OF REPERFUSION INJURY

The early restitution of blood flow to an ischemic tissue is essential to halt the progression of cellular injury associated with decreased delivery of

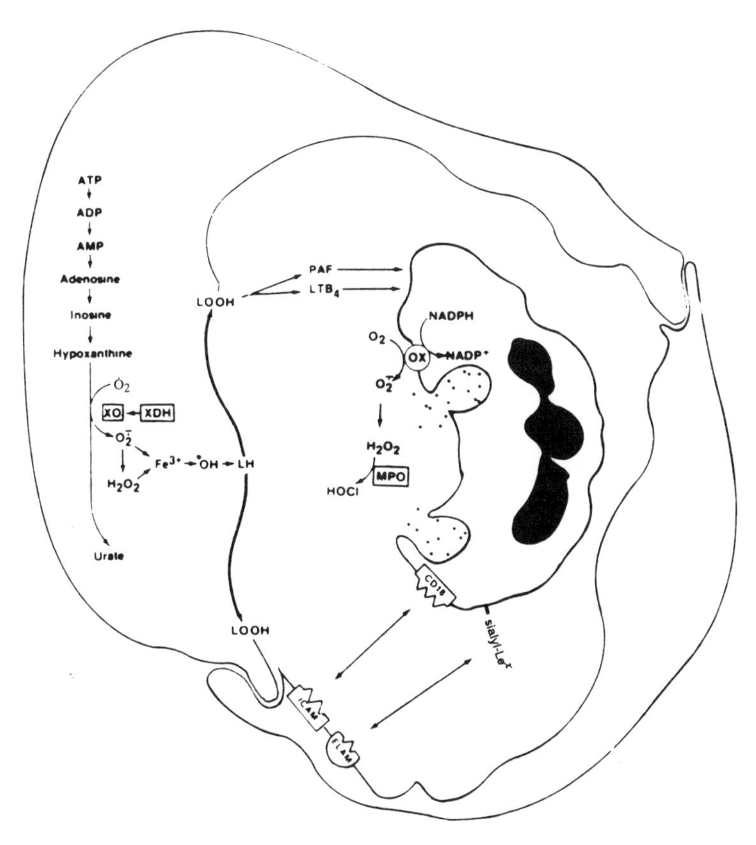

Figure 3–1 Mechanism proposed to explain the interaction of xanthine oxidase-derived oxidants, granulocyte infiltration, and microvascular dysfunction in postischemic tissues. XO = xanthine oxidase; XDH = xanthine dehydrogenase; OX = NADPH oxidase; MPO = myeloperoxidase; $\dot{O}_2$ = superoxide anion radical; H_2O_2 = hydrogen peroxide; $\dot{O}H$ = hydroxyl radical; LH = polyunsaturated fatty acid; LOOH = lipid hydroperoxide; ELAM = endothelial cell-leukocyte adhesion molecule; ICAM = intercellular adhesion molecule; sialyl-Le^x = carbohydrate ligand sialyl Lewis-X; CD18 = neutrophilic membrane glycoprotein adhesion complex; ● = neutrophil granule. (Adapted from Carden and Korthuis 1989.)

oxygen and metabolic substrates and for removal of potentially injurious metabolic by-products. Although the necessity for re-establishing an adequate blood supply is without question, there is evidence that reperfusion results in a complex series of phenomena that paradoxically injure tissues. For example, reperfusion is associated with the formation of reactive oxygen metabolites that may cause additional tissue injury. In addition, many tissues (for example, heart, skeletal muscle, brain, and liver) exhibit the no-reflow phenomenon when blood flow is reinstituted (Carden and Korthuis 1989; Schmid-Schonbein 1987). That is, some capillaries fail to reperfuse when blood flow is restored, which limits the delivery of oxygen and nutrients during reperfusion. Thus, local ischemia and hypoxia may persist during reperfusion despite the restoration of adequate arterial perfusion pressures.

Several studies dramatically illustrate the phenomenon of reperfusion injury. For example, only minor morphologic alterations are observed in the heart during ischemia (Hearse 1977). Reperfusion is, however, associated with major ultrastructural damage, cell swelling, massive cytosolic enzyme release, and prolonged depression of contractile function. In addition, reperfusion induces a number of electrophysiologic abnormalities that often produce potentially lethal arrythmias (Hearse 1977). Other evidence that suggests that reperfusion extends the injury associated with ischemia is provided by the observation that 4 h of ischemia produces far less injury to the small intestine than does 3 h of ischemia followed by 1 h of reperfusion (Parks and Granger 1986a).

Another argument that has been used to support the concept that reperfusion per se is largely responsible for the cellular dysfunction observed in ischemia/reperfusion is based on studies that compare the effectiveness of therapeutic agents administered before ischemia and after reperfusion. The results of most of these studies indicate that administration of these substances immediately before reperfusion is as effective in attenuating injury as when the agents are administered before ischemia (Carden et al. 1990a,b; Morris et al. 1987).

Recent studies indicate that reperfusion with normoxic blood is associated with microvascular and parenchymal cell injury whereas anoxic reperfusion produces minimal damage (Korthuis et al. 1989). Shifting from anoxic to normoxic blood during reperfusion is, however, associated with the production of tissue injury similar to that seen with normoxic reperfusion alone. The implication of these studies is that reperfusion initiates a series of cytotoxic events that are associated with the return of oxygenated blood. It now appears that cytotoxic events initiated on reperfusion involve the formation of reactive oxygen species that are derived from molecular oxygen.

ROLE OF REACTIVE OXYGEN METABOLITES IN POSTISCHEMIC CELLULAR INJURY

The observation that reperfusion (reoxygenation) of ischemic (hypoxic) tissues produces injury suggests that reperfusion injury may be due, at least in part, to the formation of reactive oxygen metabolites. Because the oxidants formed during reperfusion appear to be derived from molecular oxygen and most are radical species (molecules with an odd or spin-unpaired electron in their outer shell), they are often referred to as reactive oxygen metabolites or oxygen-derived free radicals. Although oxygen can accept a total of four electrons to form water, it can be reduced in univalent steps to generate three types of reactive oxygen metabolites. The addition of a single electron to the oxygen molecule results in the formation of the superoxide anion ($\dot{O}_2^-$). The two-electron reduction of oxygen results in the formation of hydrogen peroxide (H_2O_2), a reactive compound that is not a free radical but that can participate in reactions resulting in free-radical generation. The addition of a third electron results in the formation of the hydroxyl radical ($\dot{O}H$). Finally, the addition of a fourth electron to oxygen yields the nonreactive water molecule. Superoxide, hydrogen peroxide, and hydroxyl radicals have all been implicated in the genesis of postischemic cellular dysfunction.

Molecular Pathologic Effects Associated with Cellular Oxidant Production

Because the thermodynamic potential of unpaired electrons to form an electron pair is high, oxygen metabolites with unpaired electrons are highly reactive. The cumulative result of the cascade of oxidant-initiated reactions is damage to the entire array of biomolecules found in tissues, including structural, contractile, and transport proteins, enyzmes, receptors, membrane lipids, glycosaminoglycans, and nucleic acids (Halliwell and Gutteridge 1989). The molecular pathologic effects of oxidant formation include DNA "nicking" and other abnormalities, peroxidation of membrane lipid components, and crosslinking and degradation of proteins. These oxidant-induced alterations in molecular structure and organization can decimate cellular function (Figure 3-2). For example, peroxidation of linolenic acid in the acyl chains of cardiolipin, which is essential for the activity of cytochrome oxidase, may lead to a loss of function of this critical cellular enzyme (Soussi et al. 1990). Peroxidation of membrane lipids can also produce devastating effects on membrane fluidity and cell compartmentalization, leading to altered mobility of membrane receptors, impaired second messenger function, and leakage of intracellular compounds (for example, lactate dehydrogenase and creatine kinase) into the plasma. In addition, lipid peroxidation often precedes cell lysis in postischemic tissues and is associated with increased rates of protein degradation. By attacking the glycosaminoglycans and proteins in the substrata supporting intestinal epithelial cells, reactive oxygen metabolites may contribute to the exfoliation of the intestinal mucosa noted in severe ischemia/reperfusion. Oxidant-mediated damage to structural, transport, and contractile proteins and inactivation of receptors and essential enzymes in postischemic myocardium and skeletal muscle may explain the contractile dysfunction noted in these tissues during reperfusion.

Detection of Oxidant Production in Tissues Subjected to Ischemia/Reperfusion

Most of the evidence for the production of reactive oxygen metabolites during ischemia/reperfusion is indirect; it is derived primarily from the

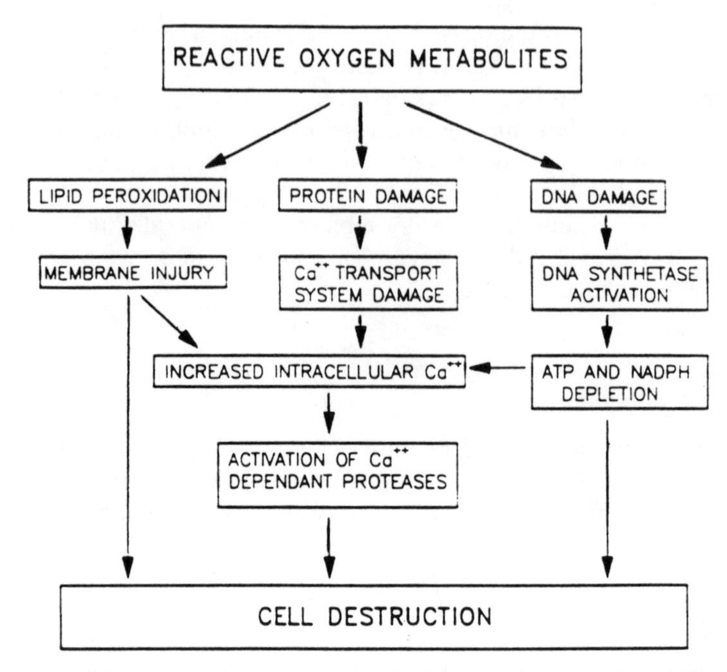

Figure 3–2 Molecular pathology associated with oxidant production. (Adapted from Carden et al. 1989.)

detection of the by-products of oxidant-induced lipid peroxidation (for example, malondialdehyde, lipid hydroperoxides, and conjugated dienes) in postischemic tissues (see Lindsay et al. 1988 for references). However, measurement of malondialdehyde and conjugated dienes can be nonspecific and may not be solely indicative of lipid peroxidation. Thus, to demonstrate definitively the production of reactive oxygen species in vivo, more direct measurements of radical production are necessary. This has been accomplished by use of electron spin resonance (ESR) spectroscopy (see Mason and Morehouse 1989 and Davies 1989 for references). Using this methodology, free radicals have been detected during reperfusion in the venous effluent draining isolated hearts and in frozen samples of myocardium obtained during reperfusion. In the latter studies, the production of radicals was largely attenuated in hearts treated with human recombinant superoxide dismutase (SOD) or the iron chelator desferal, whereas inactivated SOD had no effect. Subsequent researchers have, however, questioned the significance of the findings obtained in frozen biopsies because the tissue biopsies were subjected to severe mechanical stress during sample preparation, which may lead to artifactual oxidant production.

ESR spin trapping is another technique used to demonstrate radical production in postischemic tissues. In these studies, the spin trap acts to convert highly reactive and short-lived radical species to more stable, detectable radical adducts. Several investigators have employed this technique in the heart, and ESR signals characteristic of oxygen (hydroxyl)

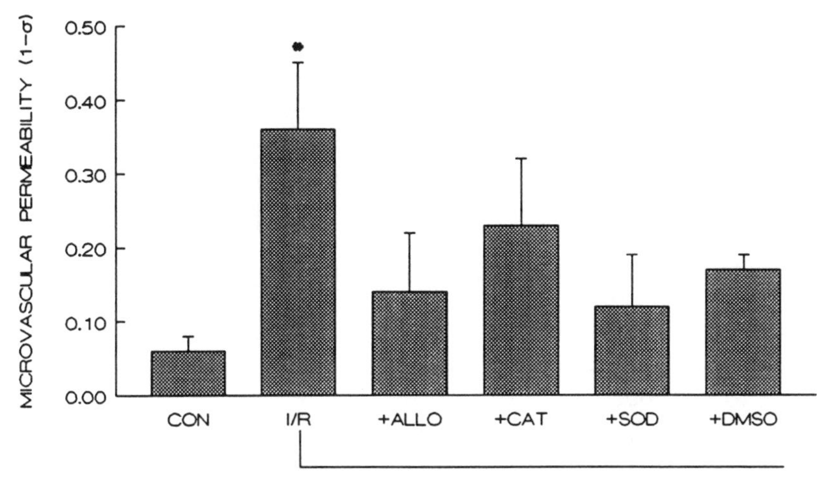

Figure 3–3 Effect of allopurinol (ALLO), catalase (CAT), superoxide dismutase (SOD), and dimethyl sulfoxide (DMSO) on the increase in microvascular permeability induced by ischemia/reperfusion (I/R) in skeletal muscle. σ represents the solvent drag reflection coefficient for total plasma proteins. An * indicates values statistically different from control. (Adapted from Korthuis et al. 1985.)

and carbon-centered radicals have been detected in coronary venous effluent, myocardial biopsy samples, and lipid extracts obtained during ischemia/reperfusion. Moreover, superoxide dismutase alone (but not peroxide-inactivated SOD), catalase, superoxide dismutase plus catalase, deferoxamine, allopurinol (an inhibitor of the oxidant-producing enzyme xanthine oxidase), and mercaptopropionyl glycine (a putative hydroxyl radical scavenger) all act to reduce the intensity of spin-trapped radical adduct signals in tissues subjected to ischemia/reperfusion. These results provide strong support for oxidant production in postischemic tissues, a notion supported by low-level chemiluminescence detection of active oxygen species during reperfusion.

Indirect Evidence Supporting a Role for Oxidant-Mediated Reperfusion Injury

The most commonly invoked evidence for the involvement of reactive oxygen metabolites in ischemia/reperfusion injury is indirect and relates to the protective effects of agents that limit the production of these cytotoxic oxidants or effectively scavenge these substances after they are produced (Figure 3-3). SOD, a specific enzymatic scavenger of superoxide, has been shown to attenuate postischemic microvascular and parenchymal cell dysfunction in many tissues (see Granger et al. 1981; Korthuis and Granger 1986; Granger 1988; Carden and Korthuis 1989 for references). Catalase, an enzyme that catalyzes the disproportionation of hydrogen peroxide to water and molecular oxygen, has also proved protective in many models of ischemia/reperfusion. Because superoxide is a relatively weak oxidant, its cytotoxic potential is generally attributed to its ability to react with hydrogen peroxide to form the highly reactive and cytotoxic hydroxyl radical. This notion is consistent with the observation that nonenzymatic scavengers of the hydroxyl radical, including dimethyl sulfoxide, dimethylthiourea, mannitol, and others, also attenuate postischemic tissue injury. Moreover, because these hydroxyl radical scavengers provide levels of protection similar to those provided by SOD and catalase, it has been suggested that this secondarily derived hydroxyl radical may play a primary role in the production of postischemic cellular dysfunction. The protective effect of metal chelators (deferoxamine, apotransferrin), which bind the transition metals necessary for formation of hydroxyl radicals, lends futher support to this notion.

Although the aforementioned studies appear to provide compelling evidence that reactive oxygen metabolites play a role in ischemia/reperfusion, the application of this notion to many organs, most notably the heart, remains controversial. The discrepant results obtained in the myocardium appear to be related to many factors including the possibility that tetrazolium staining may be an unreliable indicator of infarct size, the lack of standardized ischemic insults, and the failure to measure collateral blood flow (Reimer et al. 1989; Downey 1990). In addition, the short plasma half-lives of superoxide dismutase and catalase and the dose or manner in which these antioxidant enzymes are administered has complicated the interpretation of many studies (Turrens et al. 1984; Korthuis et al. 1991). Despite these complications, the preponderance of evidence appears to support the notion that oxidants play an important role in myocardial reperfusion injury. Indeed, when they are taken together, the direct observation of oxidant production in postischemic tissues and the ability to attenuate reperfusion-induced cellular dysfunction by antioxidant administration constitute persuasive evidence for a role for oxidants in ischemia/reperfusion.

SOURCES OF OXIDANTS IN POSTISCHEMIC TISSUES

Although there are several potential sources of cytotoxic oxidants in postischemic tissues (Table 3-1), most attention has been focused on xanthine oxidase, an enzyme found in parenchymal cells and capillary endothelium and on neutrophilic NADPH oxidase. Both enzymes are capable of producing superoxide. Hydrogen peroxide may be produced directly by xanthine oxidase or indirectly by both enzymes through the spontaneous dismutation of superoxide. Although the spontaneous dismutation of superoxide occurs rapidly under physiologic conditions, the rate of this reaction is greatly accelerated ($\times 10^4$) in the presence of the enzyme superoxide dismutase. The extremely reactive hydroxyl radical is generated through

Table 3–1 Sources of Cytotoxic Oxidants

Enzyme	Tissue	Subcellular Localization
Monoamine oxidase	Liver	Mitochondrial outer membrane
D-Amino acid oxidase	Kidney	Peroxisome
Gycolate oxidase	Liver	Peroxisome
L-Gulonolactone oxidase	Liver	Microsomal
Pyridoxamine-5′-phosphate oxidase	Liver	Cytosol
Diamine oxidase	Placenta	
Plasma amine oxidase	Plasma	
Thiol oxidase	Kidney	Plasma membrane
Urate oxidase	Liver	Peroxisome
Xanthine oxidase	Liver, intestine, endothelial cells	
Sulfite oxidase	Liver	Mitochondria
NADPH oxidase	Neutrophil	Specific granules
Auto-oxidation of catecholamines		Extracellular space
Cyclo-oxygenase		Plasma membranes

the interaction of superoxide and hydrogen peroxide in the presence of certain transition metals (for example, Fe^{3+}), which act as catalysts. Although it has been suggested that neutrophil-derived superoxide and hydrogen peroxide can interact with low-molecular-weight iron to yield hydroxyl radicals, recent data indicate that granulocytes produce minimal (if any) hydroxyl radicals in vivo (Winterbourne 1986). This finding appears to be related to the fact that myeloperoxidase (a granular enzyme released by activated granulocytes, see below) can consume virtually all of the hydrogen peroxide, thereby limiting the interaction of this oxidant with superoxide and iron.

Xanthine Oxidase as a Source of Reactive Oxygen Metabolites

Although xanthine oxidase participates in the oxidation of many endogenous and exogenous substrates, it is best known for its role as the rate-limiting enzyme in nucleic acid degradation through which all purines are channeled for terminal oxidation. However, xanthine oxidase (XO) has received a great deal of attention recently because of its ability to generate superoxide and hydrogen peroxide during the oxidation of hypoxanthine or xanthine. Whereas xanthine oxidase is capable of producing both oxidants, most of the oxygen (approximately 80%) is reduced to hydrogen peroxide under physiologic conditions (Parks and Granger 1986b). This disproportionate production of H_2O_2 is enhanced by decreased pH, increased xanthine concentration, and decreased oxygen concentration.

Xanthine oxidase normally exists in nonischemic, healthy cells, predominantly as an NAD+-dependent dehydrogenase (XDH). This form of the enzyme utilizes NAD+ instead of oxygen as the electron acceptor during oxidation of purines and does not produce superoxide or hydrogen peroxide. XDH is, however, converted to the oxidant-producing XO form (D-to-O conversion) during tissue ischemia (McCord et al. 1985; Parks and Granger 1986b). D-to-O conversion may occur through two mechanisms. The dehydrogenase can be reversibly converted to the oxidase form by oxidation or irreversibly converted following proteolysis. Although D-to-O conversion during ischemia appears to occur at different rates in different organs, the amount of D-to-O conversion is proportional to the duration of tissue ischemia. For example, XO represents approximately 19% of the total activity (XDH + XO) in rat small intestine. During ischemia this increases to 34%, 46%, and then 61% following 1, 2, and 3 hours of ischemia, respectively. D-to-O conversion in the ischemic intestine appears to occur by an irreversible proteolytic mechanism since it can be inhibited by the protease inhibitor soybean trypsin inhibitor.

Many studies have been performed in an attempt to ascertain the role of xanthine oxidase in postischemic cellular dysfunction (for example, see Parks and Granger 1986b; Granger 1988; Carden and Korthuis 1989; Smith et al. 1989; Morris et al. 1987; McKelvey et al. 1988). For example, the demonstration of D-to-O conversion in ischemic tissues is often cited as circumstantial evidence supporting the notion that xanthine oxidase may be a source of oxidants during reperfusion. Recent studies, however, have questioned the significance of D-to-O conversion in the pathogenesis of this injury. When isolated rat hepatocytes were reoxygenated after anaerobic incubation, there was not a detectable increase in xanthine oxidase yet significant hepatocellular injury was observed (deGroot and Littauer 1988). Similar findings have been noted in the rat heart (Kehrer et al. 1987). Despite the fact that anoxia/reoxygenation-induced cellular injury occurs in the absence of measurable D-to-O conversion, hepatocellular damage was still attenuated by xanthine oxidase depletion (McKelvey et al. 1988). Thus, it appears that, although D-to-O conversion may not be a prerequisite for the production of postischemic tissue injury, basal levels of xanthine oxidase may be sufficient to produce reperfusion injury in some tissues. In this regard, it is important to note that cultured endothelial cells can be severely damaged following exposure to levels of xanthine oxidase that are much lower than those measured in many nonischemic tissues if sufficient substrate is present (Bishop et al. 1985).

Despite these reservations concerning the relevance of D-to-O conversion in the pathogenesis of reperfusion injury, there is compelling evidence that supports the concept that xanthine oxidase-derived oxidants contribute to postischemic cellular dysfunction. The xanthine oxidase inhibitors allopurinol, oxypurinol, pterin aldehyde, and lodoxamide attenuate reperfusion (reoxygenation) injury in a number of different in vivo and in vitro preparations (see Smith et al. 1989; Parks and Granger 1986b; Granger 1988; Carden and Korthuis 1989; Morris et al. 1987 for references). However, the use of allopurinol and oxypurinol has been criticized because both of these agents can act as hydroxyl radical scavengers, and oxypurinol is a scavenger of neutrophil-derived hypochlorous acid (Moorhouse et al. 1987; Das et al. 1987). Inasmuch as these effects are only manifest when administered at high doses, careful attention to dosing regimes obviates this concern.

Another argument that has been used to defend the position that xanthine oxidase is an important source of oxidants in postischemic tissues is that xanthine oxidase depletion (by administration of a tungsten-supplemented, molybdenum-deficient diet prior to the ischemic insult) attenuates ischemia/reperfusion-induced increases in tissue injury (Smith et al. 1989; Granger 1988). Finally, the observation that local intra-arterial infusion of xanthine oxidase and hypoxanthine produces injury that closely mimics that produced by ischemia/reperfusion lends additional support for a role for xanthine oxidase in reperfusion injury (Granger 1988).

Because xanthine oxidase levels vary markedly from one organ to another and the enzyme is not present in some species (Parks and Granger 1986b), the xanthine oxidase hypothesis should not be extrapolated to all organs or across species without proper validation. Rabbit heart and canine skeletal muscle do not contain detectable levels of xanthine oxidase, and xanthine oxidase inhibitors do not attenuate reperfusion injury in these models (Downey 1990; Carden and Korthuis 1989). However, allopurinol reduces myocardial injury in pigs despite the fact that xanthine oxidase activity is not detected in this species. This inconsistency may be explained by the ability of allopurinol at high doses to act as an oxidant scavenger. The relevance of xanthine oxidase to the production of postischemic myocardial injury in the human has also been questioned because detectable levels of this enzyme are not present (Downey 1990). Human hearts do, however, produce uric acid during ischemia/reperfusion associated with coronary angioplasty (Huizer 1989), suggesting the presence of xanthine oxidase activity. This apparent discrepancy may be related to the fact that immunolocalization studies indicate that xanthine oxidase is found primarily in capillary endothelial cells in human myocardium (see Parks and Granger 1986b for references). Because endothelial cell xanthine oxidase activity is low relative to that reported in many other cell types and because endothelial cells probably account for $<1\%$ of myocardial weight, only low levels of xanthine oxidase would be present in homogenates of human myocardium. Additional studies are clearly needed to resolve the controversy regarding the role of xanthine oxidase in human myocardial reperfusion injury.

Neutrophilic NADPH Oxidase as a Source of Oxidants

Polymorphonuclear leukocytes are armed with an impressive arsenal of bactericidal agents that allow these cells to play a vital role in host defense against invading pathogens. These same agents can produce extensive cellular damage in host tissues when leukocytes are activated during inflammatory conditions such as ischemia/reperfusion (Weiss 1989). For example, phagocytic leukocytes contain a membrane-bound NADPH oxidase that is activated when neutrophils are exposed to chemoattractants, immune stimuli, or bacteria. This enzyme oxidizes cytoplasmic NADPH to $NADP^+$ and generates two molecules of superoxide that can subsequently dismutate to form hydrogen peroxide. Activated neutrophils also secrete myeloperoxidase (MPO), an enzyme that catalyzes the formation of hypochlorous acid (HOCl) from hydrogen peroxide and chloride ions. As noted previously, the hydrogen peroxide that fuels this reaction is derived from the spontaneous dismutation of superoxide. HOCl is a potent oxidizing and chlorinating agent that reacts rapidly with primary amines to yield cytotoxic N-chloramines, which contain the two oxidizing equivalents of hydrogen peroxide and HOCl. The cellular toxicity associated with HOCl and N-chloramines appears to depend on their relative lipophilicities and probably occurs as a result of sulfhydryl oxidation, hemeprotein and cytochrome inactivation, chlorination of purine bases on DNA, and degradation of amino acids and proteins. In addition, neutrophil-derived HOCl and hydrogen peroxide can also initiate hydrolytic reactions catalyzed by neutrophilic proteases. For example, HOCl can alter the balance between tissue proteases and antiproteases either by oxidative inactivation of several plasma protease inhibitors (for example, alpha-1-proteinase inhibitor) or by activation of latent proteases (for example, collagenase or gelatinase) or both. Several reports indicate that neutrophil-derived proteases may play an important role in the cellular dysfunction induced by ischemia/reperfusion (see Inauen et al. 1990 for references). Thus, it is apparent that the neutrophil utilizes the NADPH oxidase system and granular constituents in a coordinated fashion to facilitate the destructive potential of the granulocyte.

ROLE OF GRANULOCYTES IN POSTISCHEMIC CELLULAR DYSFUNCTION

Accumulation of Neutrophils in Postischemic Tissues

Several approaches have been used to define the responses and role of leukocytes in postischemic tissue injury. One approach involves monitoring the appearance of radiolabelled leukocytes or granulocyte-specific enzymes, such as myeloperoxidase, in postischemic tissues. Data derived using this technique indicate that reperfusion of ischemic tissues is associated with massive leukocyte infiltration (see Granger 1988; Carden and Korthuis 1989 for references). Studies in the small bowel and skeletal muscle have demonstrated that the neutrophil accumulation produced by reperfusion can be effectively attenuated by pretreatment with a variety of oxidant scavengers (for example, superoxide dismutase, catalase, dimethylthiourea, and deferoxamine). These observations suggest that reactive oxygen metabolites play a role in the recruitment of neutrophils into postischemic tissues (Granger 1988; Carden and Korthuis 1989). Because

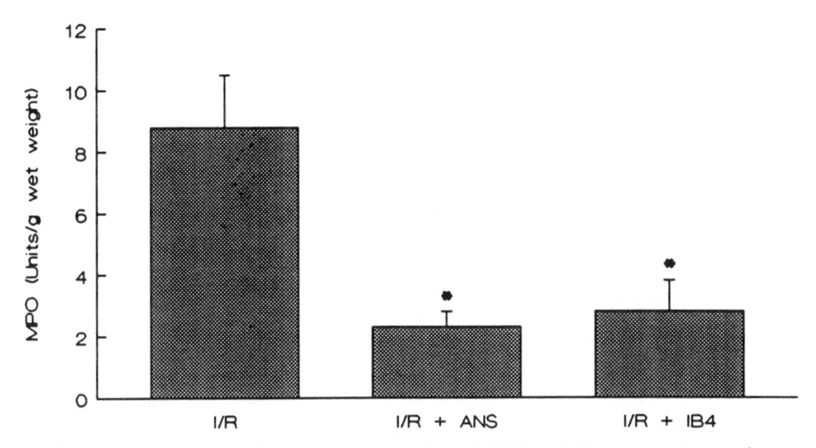

Figure 3–4 Alterations in tissue myeloperoxidase (MPO) activity, a sensitive indicator of neutrophil infiltration, after 4 h of inflow occlusion (I/R), I/R + antineutrophil serum (ANS), and I/R + inhibition of neutrophil adherence with a monoclonal antibody (IB_4) directed against CD_{11}/CD_{18} in skeletal muscle. An * indicates values statistically different from control (CON). (Adapted from Carden et al. 1990b.)

allopurinol is as effective as the oxidant scavengers in attenuating reperfusion-induced leukocyte accumulation, xanthine oxidase appears to be a source of the oxidants involved in leukocyte recruitment (Figure 3-4; see also Figure 3-1).

Effect of Leukopenia or Neutropenia on the Cellular Dysfunction Associated with Ischemia/Reperfusion

An important question that arises from the observations of leukocyte infiltration into postischemic tissues is whether leukocyte accumulation and activation is a cause or an effect of reperfusion injury. To address this issue, animals have been rendered leukopenic by use of Leukopak filters (Baxter-Travenol, San Francisco), by administration of antitumor agents such as hydroxyurea or by exposure to x-irradiation. Leukopenia greatly attenuates postischemic microvascular and parenchymal cell dysfunction in a variety of tissues (see Granger 1988; Carden and Korthuis 1989 for references). Although the aforementioned results support the concept that leukocytes play a role in the pathogenesis of postischemic reperfusion injury, these studies suffer from the relative nonspecificity of the interventions employed to produce neutrophil depletion. For example, use of Leukopak filters to deplete neutrophils also effectively depletes circulating platelets (Carden et al. 1990a,b). Platelets have been implicated in the development of the no-reflow phenomenon, and platelet-derived substances can increase microvascular permeability. Thus, the results obtained in studies in which Leukopak filters are employed to produce neutropenia must be interpreted with caution. Similarly, because administration of the antitumor agents and exposure to x-irradiation reduce all types of leuko-

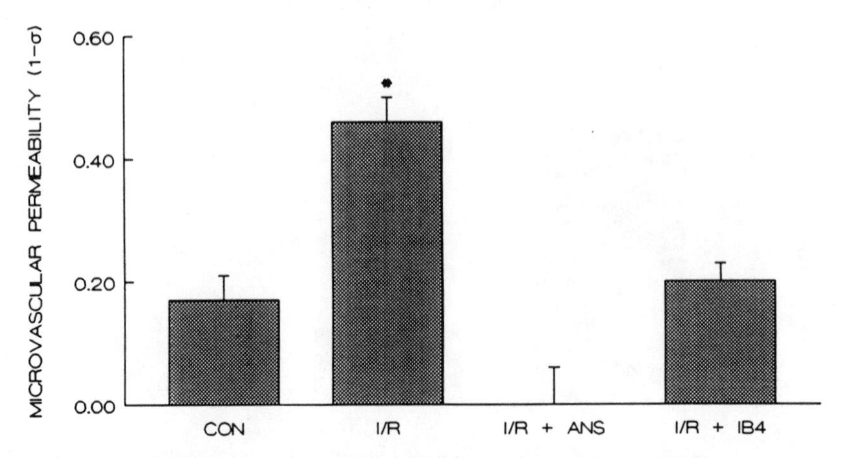

Figure 3–5 Effect of administration of antineutrophil serum (ANS) or a monoclonal antibody (IB_4) directed against the neutrophilic membrane glycoprotein CD_{11}/CD_{18} adherence complex on the increase in microvascular permeability induced by ischemia/reperfusion (I/R) in skeletal muscle. σ represents the solvent drag reflection coefficient for total plasma proteins. An * indicates values statistically different from control (CON). (Adapted from Carden et al. 1990b.)

cytes (granulocytes, lymphocytes, monocytes), it is difficult to determine which cell line participates in the genesis of ischemia/reperfusion injury. Moreover, the generalized effects associated with intravenous administration of the antitumor agents and whole body irradation raise the possibility that some as yet unrecognized effect of these pertubations contributed to the attenuation of injury.

More definitive evidence for a role of neutrophils in postischemic cellular dysfunction is provided by the observation that neutropenia produced by administration of highly specific antineutrophil serum attenuates the extent of reperfusion injury in the heart, skeletal muscle, stomach, small bowel, skin, and kidneys (for example, Figure 3-5; see McCord 1988; Granger 1988; Carden et al. 1990a,b for references). These findings indicate that granulocyte infiltration into postischemic tissues is a cause, rather than an effect, of ischemia/reperfusion. However, this interpretation is complicated by the fact that the altered chemical composition of extracellular fluid in postischemic tissues may lead to suppression of granulocyte function. Neutrophilic superoxide production is reduced by hypoxia, acidosis, and adenosine. Tissue PO_2 is likely to remain low in areas of tissues exhibiting the no-reflow phenomenon, thereby limiting the availability of this substrate for the neutrophilic oxidative burst. Interstitial pH during ischemia may fall to as low as 6.0, a level which suppresses neutrophilic superoxide production by 80% in vitro. Finally, adenosine, which accumulates in ischemic tissues, inhibits neutrophilic superoxide production and adherence of granulocytes to microvascular endothelium (Grisham and Granger, 1989). Indeed, exogenous administration of adenosine limits the extent of reperfusion injury in heart, skeletal muscle, and small bowel.

Role of Neutrophil Adherence in Postischemic Tissue Injury

For a neutrophil to produce microvascular and parenchymal cell injury, it must establish close apposition with the target cell. Thus, increased neutrophil adhesiveness for microvascular endothelium is a critical first step in the sequence of events leading to granulocyte-dependent microvascular and parenchymal injury. Granulocyte accumulation in the interstitial spaces is a characteristic feature of acute inflammatory conditions such as ischemia/reperfusion. The egress of leukocytes from the vascular to extravascular compartment also requires that the granulocyte first make intimate contact with the endothelial cells. In vitro studies suggest that radial dispersive forces are involved in leukocyte rolling that occurs in postcapillary venules. That is, as blood vessel diameter increases in moving from the capillaries to postcapillary venules, the more flexible erythrocytes begin to pass the leukocyte, which, in turn, deflects the white cells toward the vessel walls and margination occurs. By this mechanism, nearly all leukocytes exiting capillaries are forced to marginate in the postcapillary venules (Schmid-Schonbein et al. 1980). Whether leukocytes adhere to venular endothelium depends largely on the balance between adhesion forces generated by the leukocyte and endothelium and the hydrodynamic dispersal forces (for example, blood flow velocity) that tend to sweep leukocytes away from the vascular wall. Thus, for a leukocyte to remain stationary within the vasculature, it must bind to endothelium with sufficient strength to withstand the shear rates and associated forces acting to sweep it away. Because shear rates tend to be lower in venules than in arterioles and radial dispersive forces tend to establish margination in venules, the propensity for leukocyte/endothelial cell adhesive forces to overcome hydrodynamic dispersal forces is considered to be much greater in postcapillary than precapillary vessels. For the same reason, ischemia facilitates granulocyte adhesion to venular endothelium.

An important factor that modulates the adhesive interaction between granulocytes and the venular endothelium is the postadhesive force generated by adhesion molecules expressed on the surface of neutrophils and endothelial cells (Albelda and Buck 1990; Springer et al. 1989; Springer 1990; Kishimoto et al. 1990; Jutila et al. 1989). A variety of proinflammatory mediators (for example, formylated peptides such as FMLP, platelet-activating factor (PAF), leukotriene B_4 (LTB_4), and the complement component (C5a) are known to act directly on the neutrophil to enhance its adherence to endothelium. The proadhesive action of these mediators occurs rapidly (within 1 to 2 min) and is dependent on the mobilization of specific adhesion molecules to the surface of the leukocyte. Most available evidence implicates the leukocyte glycoprotein adherence complex designated CD_{11}/CD_{18} (or Mac-1, LFA-1, p150,95) as the primary mediator of adhesion of activated neutrophils to endothelium. This adhesive glycoprotein complex is composed of a family of three structurally and functionally related glycoprotein heterodimers, each consisting of an immunologically distinct α-subunit (CD_{11}) that is noncovalently associated with a common β-subunit (CD_{18}). LFA-1 (CD_{11a}/CD_{18}) is present in all leukocytes, whereas Mac-1 (CD_{11b}/CD_{18}) and p150,95 (CD_{11c}/CD_{18}) are expressed on granulocytes and monocytes. Mac-1 (CD_{11b}/CD_{18}) is rapidly upregulated on the surface of neutrophils exposed to proinflammatory agents such as C5a as well as in tissues exposed to ischemia/reperfusion.

Another type of adhesive molecule associated with the granulocyte is the MEL-14 antibody-defined antigen designated $gp100^{MEL\text{-}14}$ (also LECAM1) (Albeda and Buck 1990; Springer 1990, Jutila et al. 1989). The MEL-14 antigen modulates the adhesive interaction between unstimulated neutrophils and vascular endothelium and is rapidly downregulated when granulocytes are exposed to C5a, phorbol esters, and LTB_4. Release of $gp90^{MEL\text{-}14}$ from the surface of the neutrophil appears to account for this downregulation. The release occurs rapidly (within 1 to 5 min) and is associated with loss of the ability of these cells to home to inflammatory sites (Jutila et al. 1989). The functional significance of these observations is that the neutrophil MEL-14 antigen may mediate the initial adhesive interaction between granulocytes and endothelium at sites of inflammation. When they are activated by chemotactic signals, the neutrophils at the inflamed site downregulate the expression of the MEL-14 antigen, thereby allowing the neutrophil to proceed from binding to the endothelium to extravasating into the tissues. In addition, this mechanism may prevent activated granulocytes, which dislodge from reperfused regions (due to elevated shear forces associated with vasodilation), from extravasating into and damaging healthy tissues at sites distant from the inflammatory locus.

Microvascular endothelial cells also express molecules that modulate their adhesive interaction with leukocytes (Springer et al. 1989; Springer 1990; Bevilacqua et al. 1989). Intercellular adhesion molecule-1 (ICAM-1) and endothelial leukocyte adhesion molecule-1 (ELAM-1 or LECAM2) are inducible surface proteins that are present on endothelial cells following inflammation or following exposure to cytokines. Although both of these proteins modulate the adhesive interaction between leukocytes and endothelium, ELAM-1 is distinguished from ICAM-1 by its lack of expression is unstimulated cells. LFA-1 (CD_{11a}/CD_{18}) appears to act as a ligand for ICAM-1 in unstimulated neutrophils. However, when neutrophils are activated by a proinflammatory stimulus such as FMLP, Mac-1 (CD_{11b}/CD_{18}) appears to act as the predominant ligand for ICAM-1. ELAM-1 contains a lectin motif that recognizes a carbohydrate ligand designated sialyl-Lewis X found on glycoproteins and glycolipid carbohydrate groups on the cell surface of neutrophils (Phillips et al. 1990).

Another adhesive protein that is expressed by endothelial cells is granule membrane protein-140 (GMP-140, LECAM3, PADGEM protein or CD62) (Albelda and Buck 1989). GMP-140 is an integral membrane glycoprotein found in the secretory (α) granules of platelets and in Weibel-Palade bodies of venular endothelial cells. When endothelial cells or platelets are activated by agonists such as thrombin, GMP-140 is rapidly (within seconds

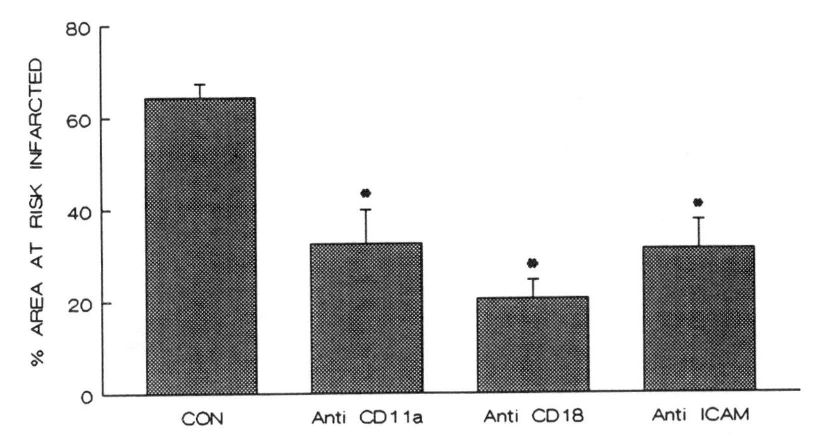

Figure 3–6 Reduction in percent area at risk for infarction with monoclonal antibodies against intercellular adhesion molecule-1 (ICAM) expressed by endothelial cells and the α-chain CD_{11a} and common β-chain CD_{18} of the neutrophil membrane glycoprotein adherence complex designated CD_{11}/CD_{18}. An * indicates values statistically different from control (CON). (Adapted from Seewaldt-Becker et al. 1989.)

to minutes) mobilized to the cell surface where it may participate in the modulation of leukocyte-endothelial cell interactions. The ligand for GMP-140 appears to be a pentasaccharide (lacto-N-fucopentaose III) that contains the nonsialylated version of the Lewis X epitope (Brandley et al. (1990).

Monoclonal antibodies directed against function-related epitopes on some of the adhesion molecules have been used to probe the importance of endothelial cell-leukocyte interactions in the production of postischemic injury to identify which specific adhesion molecules mediate this interaction. Preventing neutrophil adherence by administration of monoclonal antibodies (MoAb 60.3 and IB_4) directed against the common β chain (CD_{18}) before ischemia or at the onset of reperfusion largely attenuates ischemia/reperfusion-induced neutrophil infiltration (see Figure 3-4), anoxia/reoxygenation injury in endothelial cell monolayers, postischemic intestinal and skeletal muscle vascular dysfunction, ischemia/reperfusion-induced increases in myocardial infarct size, and parenchymal cell injury in postischemic skin (for example, see Figure 3-5 and Granger 1988; Carden et al. 1990a,b; Lucchesi 1990; Springer et al. 1989; Vedder et al. 1990). In addition, antibodies against LFA-1 (CD_{11a}/CD_{18}) and Mac-1 ($CD_{11b}CD_{18}$) reduce myocardial infarct size in rabbit and dog hearts subjected to ischemia/reperfusion (Figure 3-6). These studies lend considerable support to the notion that granulocytes play an important role in the production of postischemic cellular dysfunction. It also appears that activated neutrophils adhere to parenchymal cells by a CD_{11}/CD_{18}-dependent mechanism, a phenomenon associated with increased oxidant production (Entman et al. 1990). Finally, inhibition of granulocyte adherence with MoAb IB_4 prevents the postischemic increase in skeletal muscle vascular resistance,

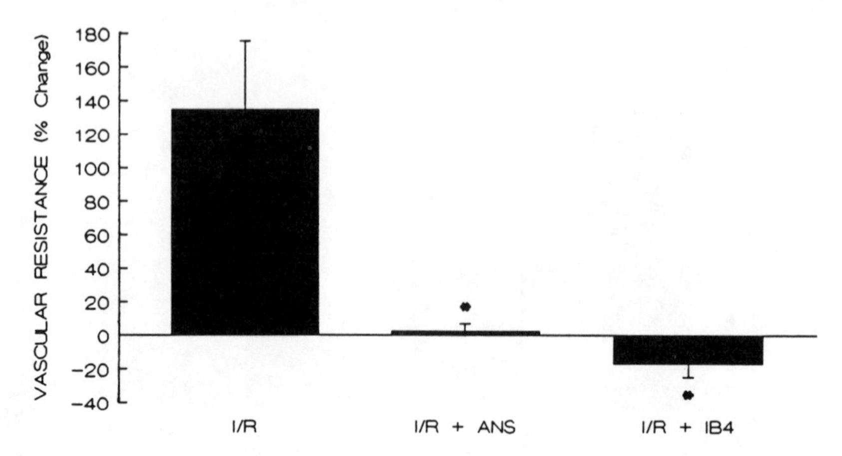

Figure 3–7 Changes in vascular resistance induced by ischemia reperfusion (I/R) alone, I/R + antineutrophil serum (ANS), and I/R + monoclonal antibody (IB_4) directed against the CD_{11}/CD_{18} complex. An * indicates values statistically different from control. (Adapted from Carden et al. 1990b.)

which suggests that neutrophil adherence to the microvascular endothelium plays an essential role in the development of the no-reflow phenomenon (Figure 3-7) (Carden et al. 1990a,b). The importance of other adhesion molecules expressed by neutrophils (for example, CD_{11c}/CD_{18} and $gp100^{MEL-14}$) has not been established.

Minimal information is available regarding the role of adhesion molecules expressed by endothelial cells in modulating the adhesive interaction between granulocytes and endothelial cells in tissues subjected to ischemia/reperfusion. Recent reports indicate, however, that anti-ICAM-1 antibodies reduce myocardial infarct size in rabbit hearts subjected to ischemia/reperfusion (see Figure 3-6; Seewaldt-Becker et al. 1990), attenuate postischemic neutrophil adherence in mesenteric venules (Granger, unpublished observations), and limit anoxia/reoxygenation-induced endothelial cell dysfunction (Kvietys, unpublished observations). These observations support the notion that adhesion molecules expressed by endothelial cells also act to modulate the leukocyte/endothelial interactions in postischemic tissues. Although it has been suggested that ELAM-1 and GMP-140 may play a role ischemia/reperfusion-induced leukocyte/endothelial interactions and cellular dysfunction, these issues have not been specifically addressed.

The observation that granulocyte depletion and prevention of granulocyte adherence are equally effective in attenuating reperfusion injury suggests that adherence of granulocytes to microvascular endothelium is a limiting factor in ischemia/reperfusion-induced cellular dysfunction. However, the chemical mediators of this dysfunction remain uncertain. Since oxy-radical scavengers also attenuate ischemia/reperfusion injury, it is tempting to attribute the injury process entirely to oxidants. This observation alone does not, however, constitute definitive support for this idea because SOD and catalase also prevent neutrophil adherence, indicating that oxidants may function primarily to recruit granulocytes into postischemic tissues (Inauen et al. 1990). In addition to releasing oxidants, neutrophils also release a variety of hydrolytic enzymes (for example, collagenase, elastase, cathepsin G), which are capable of producing cellular dysfunction. Indeed, inhibition of neutrophilic elastase is as effective as the oxy-radical scavengers in attenuating ischemia/reperfusion-induced increases in microvascular permeability. There is also evidence indicating that oxidant production is required for some neutrophilic proteases to produce tissue injury (Weiss et al. 1989). For example, collagenase activation requires hypochlorous acid.

There are several mechanisms whereby neutrophils may induce injury in tissues subjected to ischemia/reperfusion. Because granulocytes are relatively large, stiff, viscoelastic cells that become even less deformable when exposed to conditions associated with ischemia (for example, hypoxia, decreased pH), they can lodge within and occlude microvessels (Schmid-Schonbein 1987). This phenomenon is exacerbated by ischemia-induced endothelial cell swelling and by the increased adhesive interaction between leukocytes and endothelium associated with expression of CD_{11}/CD_{18}, ICAMs, and possibly ELAMs, GMP-140, and $gp90^{MEL-14}$ (Carden et al. 1990a,b). These changes may all contribute to the development of the no-reflow phenomenon (Figure 3-8). Thus, the protection afforded by neutrophil depletion or by preventing leukocyte adherence may result from improved tissue perfusion during ischemia/reperfusion. A second important consequence of the adhesive interaction between neutrophils and endothelial cells is the establishment of a "protected microenvironment" in the contact area between these cells (Weiss et al. 1989). Because it protects the neutrophil-derived oxidants and proteases from the effects of circulating antioxidants and antiproteases, the formation of this area of close contact facilitates the destructive potential of granulocyte. Moreover, granulocyte adherence to target cells appears to enhance oxidant production by stimulated neutrophils (Entman et al. 1990).

Role of Oxidants in the Elaboration of Chemotactic Factors in Postischemic Tissues

Figure 3-9 illustrates how a diapedesing neutrophil moves along the chemoattractant concentration gradient to the parenchymal cell, where it unleashes its cytotoxic arsenal and produces tissue injury. Although several proinflammatory mediators have been proposed to explain the granulocyte infiltration associated with reperfusion of ischemic tissues, the nature of this chemotactic signal remains unresolved. However, the observation that xanthine oxidase inhibition, superoxide dismutase, or neutrophil depletion prevents both the influx of neutrophils and the cellular dysfunction associated with reperfusion of ischemic bowel suggests that ischemia/reperfusion results in xanthine oxidase-generated, superoxide-dependent elaboration of proinflammatory agents that attract and activate inflammatory

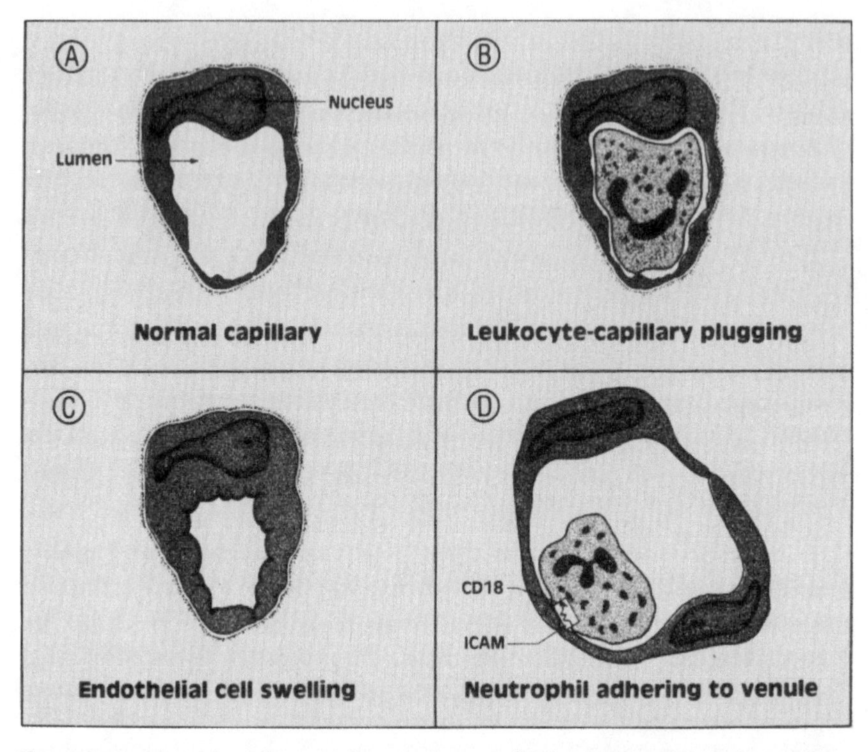

Figure 3–8 Potential mechanisms that may account for the development of the no-reflow phenomenon. **A**. Normal capillary. **B**. Leukocyte plugging a microvessel may be related to increased white cell stiffness and ischemic swelling of leukocytes and endothelial cells. **C**. Endothelial cell swelling reduces vascular caliber leading to increased resistance. **D**. Leukocyte adherence in postcapillary venules decreases the diameter of the vascular lumen and increases vascular resistance. ICAM refers to intercellular adhesion molecules that are expressed by endothelial cells and CD18 refers to adhesive glycoprotein complex expressed by activated neutrophils. (Adapted from Carden et al. 1990b.)

neutrophils to the mucosa of the small bowel, where neutrophil-derived oxidants or proteases or both exacerbate intestinal injury (Granger 1988; Inauen et al. 1989). Subsequent studies employing intravital microscopic approaches have demonstrated that allopurinol and superoxide dismutase attenuate both leukocyte adherence and extravasation in the postischemic mesentery (Figure 3-10; Inauen et al. 1989). Superoxide dismutase also decreases hypoxia/reoxygenation-induced neutrophil adherence to endothelial cell monolayers. Although SOD reduces the adherence of activated neutrophils to endothelial cells, it does not influence the ability of activated neutrophils to adhere to a biologically inert surface, a result that suggests that endothelial cells are required for superoxide-mediated leukocyte adherence.

Generation of leukotriene B_4 (LTB_4), a potent chemoattractant for neutrophils, may also play a role in the ischemia/reperfusion-induced chemotactic response in some tissues. For example, mucosal LTB_4 levels and

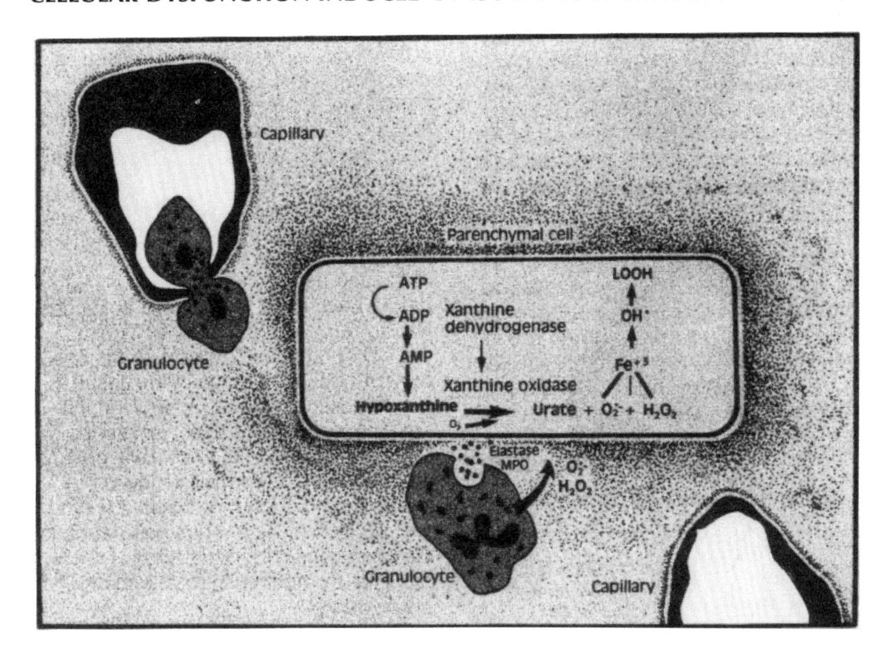

Figure 3–9 Parenchymal cell injury induced by activated granulocytes. Diapedesing granulocyte moves out of capillary in upper left portion of the figure and moves along concentration gradient of chemoattractants produced as a result of formation of xanthine oxidase-generated oxidants in parenchymal cell. The extravasated and activated neutrophil produces tissue injury by release of oxidants (for example, superoxide, hydrogen peroxide, hypochlorous acid) and enzymes (for example, MPO, elastase). (Adapted from Carden et al. 1990b.)

intestinal myeloperoxidase activity increase significantly during reperfusion. The postischemic neutrophil infiltration and increase in LTB_4 levels are dramatically attenuated by administration of lipoxygenase inhibitors or a LTB_4 antagonist (Zimmerman et al. 1990). LTB_4 may be only one of many proinflammatory mediators formed at the onset of reperfusion that attract and activate neutrophils. Activated complement components, lipid hydroperoxides, and superoxide-dependent chemoattractants also may be involved in the granulocyte infiltration associated with reperfusion of ischemic tissues (Carden and Korthuis 1989). In addition, exposing endothelial cells to hydrogen peroxide increases the production of PAF (Lewis et al. 1988), a potent chemotactic agent that increases in many tissues during reperfusion. Inasmuch as superoxide dismutase attenuates PAF-induced increases in leukocyte adherence and PAF increases markedly in the postischemic intestine, it seems likely that this proinflammatory stimulus may play an important role in the neutrophil infiltration and cellular dysfunction produced by ischemia/reperfusion (Kubes et al. 1990). Tumor necrosis factor also possesses proinflammatory properties and may be important in interacting with PAF to induce neutrophil-dependent cellular dysfunction.

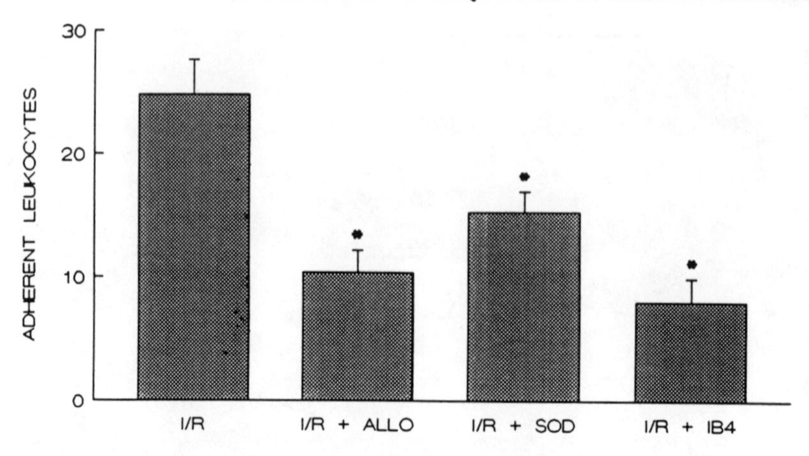

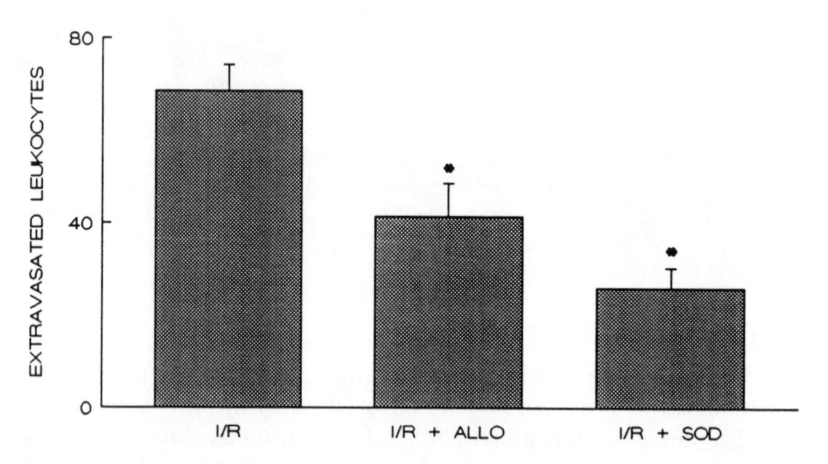

Figure 3–10 Allopurinol (ALLO) and superoxide dismutase (SOD) attenuate granulocyte adherence and extravasation in postischemic mesentery. Adherent leukocytes refers to the number of adherent cells per 100-μm length of venule and extravasated leukocytes refers to the number of leukocytes per microvascular field. I/R = ischemia/reperfusion; an * indicates values statistically different from control. (Adapted from Granger et al. 1989 and Suzuki et al. 1989.)

SUMMARY AND FUTURE DIRECTIONS

There is substantial evidence that supports the concept that xanthine oxidase-derived oxidants, produced at the time of reperfusion, initiate the formation and release of proinflammatory agents, which subsequently attract and activate neutrophils (see Figure 3-1). The activated granulocytes adhere to the vascular endothelium, extravasate, and release cytotoxic oxidants and nonoxidative toxins (for example, proteases) that contribute to endothelial cell membrane dysfunction and increase microvascular permeability (see Figure 3-1). In addition, the adherent neutrophils increase the resistance to blood flow and thus contribute to the development of the no-reflow phenomenon (see Figure 3-1). This scheme suggests that oxidant scavengers or agents that interfere with neutrophilic function may provide important new therapeutic avenues for the treatment of reperfusion-induced cellular dysfunction. Immunoneutralization of granulocyte or endothelial cell-associated adherence molecules appears to hold particular promise in this regard.

Although a large volume of evidence has accumulated over the past 10 years that supports the hypothesis presented in Figure 3-1, there are still many unanswered questions regarding this scheme. For example, we are only now beginning to unravel the role of oxidants and proteases in the postischemic adherence process. Moreover, the nature and identity of the chemotactic signal involved in ischemia/reperfusion-induced granulocyte infiltration remains uncertain. In addition, more precise definition of the role of different adhesion molecules in the modulation of ischemia/reperfusion-induced endothelial cell-leukocyte interactions may uncover important new avenues for therapeutic intervention. The function of the vascular adhesive receptors may be much more complicated than is generally appreciated. For example, in addition to modulating leukocyte/endothelial cell interactions, GMP-140 also appears to mediate adhesion of stimulated platelets to granulocytes, an association noted in a variety of inflammatory states (Hamburger and McEver 1990). This observation may have several important functional consequences. GMP-140 may serve to facilitate hemostatic and inflammatory responses in vascular injury. In this regard, it is of interest to note the GMP is localized to Weibel-Palade bodies where the hemostatic protein von Willebrand factor is concentrated. In addition, the interaction of platelets with mononuclear phagocytes has been noted early in the course of atherosclerotic plaque formation, which may ultimately result in tissue ischemia.

It is also not clear whether white cells other than granulocytes contribute to postischemic cellular dysfunction or whether the interaction of formed elements (for example, platelets and neutrophils) plays a role in the production of reperfusion injury. These may be important considerations because platelets may contribute to the development of the no-reflow phenomenon. These cells are also known to interact with both circulating neutrophils and the vascular endothelium to modulate inflammatory responses. For example, each cell type can utilize lipid intermediates (for example, leukotriene A_4) derived from the other to synthesize new products (for example, leukotriene C_4, PAF), some of which are proinflammatory (MacLouf and Murphy 1988). Other platelet-derived products may act to inhibit neutrophilic superoxide production (for example, adenine nucleotides, platelet-derived growth factor), which, under certain conditions, may afford a cellular mechanism to protect against oxidant-mediated damage (Cronstein et al. 1990). In addition, platelets may, under certain conditions, exacerbate neutrophil-dependent injury, an effect that appears to be mediated by the release of platelet-driven serotonin and enhancement of leukocyte function (Boogaerts et al. 1982).

Little is known about the potential interaction of proinflammatory mediators and tissue xanthine oxidase activity. However, recent evidence supports the concept that histamine may augment xanthine oxidase activity by an as yet undetermined mechanism (Friedl et al. 1989a,b). In addition, conversion of xanthine dehydrogenase to the oxidant-producing xanthine oxidase form occurs in endothelial cells exposed to activated granulocytes (Friedl et al. 1989a,b). The finding that neutrophil-dependent lysis of cultured endothelial cells can be inhibited by xanthine oxidase inhibition suggests that stimulated granulocytes may induce oxidant-mediated injury by activation of endothelial cell xanthine oxidase. The relevance of this in vitro finding to intact organs is, however, unclear because xanthine oxidase inhibition or inactivation does not prevent the increase in microvascular permeability produced by activated neutrophils in vivo (Smith et al. 1991; Carden et al. 1991). This apparent discrepancy may be explained by the possibility that xanthine oxidase inhibition limits the extent of endothelial cell lysis produced by activated neutrophils but fails to maintain microvascular barrier function of the endothelium.

The lack of specificity of most oxidant scavengers and the paucity of direct evidence for oxidant production during reperfusion have hampered progress in this area. For example, the protective actions of DMSO in reperfusion injury is often attributed to its ability to scavange hydroxyl radicals. However, this compound also interferes with granulocyte superoxide production, adherence, chemotaxis, and arachidonate metabolism (see Sekizuka et al. 1989 for references). The lack of direct evidence for oxidant production in vivo probably relates to the fact that few laboratories have the necessary technology (ESR spectroscopy, low-level chemiluminescence) and expertise to assess directly oxidant production in intact animals. This problem is compounded by the controversial nature of the few in vivo measurements that have been obtained with such techniques. Progress in this field will depend on the development of new techniques for monitoring oxidant production and granulocyte function in vivo and the identification of new compounds that selectively interfere with one or both of these processes. A promising development in this regard relates to a recently described technique that combines high-speed video imaging techniques with luminol-enhanced, low-level chemiluminscence to study oxidant production by activated granulocytes in vivo (Suematsu et al. 1989).

ACKNOWLEDGMENTS

Drs. Korthuis and Carden are the recipients of an Established Investigatorship from the American Heart Association (880197) and a National Research Service Award (AR-08075) from the National Heart, Lung, and Blood Institute, respectively. This work was supported by grants from the National Heart, Lung, and Blood Institute (HL-36069) and the American Heart Association (880818).

REFERENCES

Albelda, S. M., and Buck, C. A. (1990). Integrins and other cell adhesion molecules. *FASEB J.*, 4:2868–80.

Bevilacqua, M. P., Stengalin, S., Gimbrone, M. A., and Seed, B. (1989). Endothelial cell leukocyte adhesion molecule 1: an inducible receptor for neutrophils related to complement regulatory protein and lectins. *Science*, 243:1160–65.

Bishop, C. T., Mirza, Z., Crapo, J. D., and Freeman, B. A. (1985). Free radical damage to cultured porcine aortic endothelial cells and lung fibroblasts: modulation by culture conditions. *In Vitro Cell. Dev. Biol.*, 21:229–36.

Boogaerts, M. A., Yamada, O., Jacob, H. S., and Moldow, C. F. (1982). Enhancement of granulocyte-endothelial cell adherence and granulocyte-induced cytotoxicity by platelet release products. *Proc. Natl. Acad. Sci.*, 79:7019–23.

Brandley, B. K., Swiedler, S. J., and Robbins, P. W. (1990). Carbohydrate ligands on the LEC Cell Adhesion Molecules. *Cell*, 63:861–63.

Carden, D. L., and Korthuis, R. J. (1989). Mechanisms of postischemic vascular dysfunction in skeletal muscle: Implications for therapeutic intervention. *Microcirc. Endothelium Lymphatics*, 5:277–98.

Carden, D. L., Korthuis, R. J., and Granger, D. N. (1990a). Free radicals and oxygen toxicity: Mechanisms of oxidant-mediated reperfusion injury. *Crit. Care Rep.*, 1:276–81.

Carden, D. L., Smith, J. K., Zimmerman, B. J., Korthvis, R. J., and Granger, D. N. (1989). Reperfusion injury following circulatory collapse: The role of reactive oxygen metabolites. J Crit Care 4:294–307.

Carden, D. L., Smith, J. K., and Korthuis, R. J. (1990b). Neutrophil-mediated microvascular dysfunction in postischemic canine skeletal muscle: Role of granulocyte adherence. *Circ. Res.*, 66:1436–44.

Carden, D. L., Smith, J. K., and Korthuis, R. J. (1991). Oxygen-mediated, CD_{18}-dependent microvascular dysfunction induced by complement-activated granulocytes. *Am. J. Physiol.*, 260:H1144–52.

Cronstein, B. N., Daguma, L., Nichols, D., Hutchison, A. J., and Williams M. (1990). The adenosine/neturophil paradox resolved: human neutrophils possess both A_1 and A_2 receptors that promote chemotaxis and inhibit O_2 production, respectively. *J. Clin. Invest.*, 85:1150–57.

Das, D. K., Engleman, R. M., Clement, R., Otani, H., Prasad, M. R., and Rao, P. S. (1987). Role of xanthine oxidase inhibitors as free radical scavenger: a novel mechanism of action of allopurinol and oxypurinol in myocardial salvage. *Biochem. Biophys. Res. Commun.*, 148:314–19.

Davies, M. J. (1989). Direct detection of radical production in the ischaemic and reperfused myocardium: current status: *Free Radic. Res. Commun.*, 7:275–84.

de Groot, H., and Littauer, A. (1988). Reoxygenation injury in isolated hepatocytes: cell death precedes conversion of xanthine dehydrogenase to xanthine oxidase. *Biochem. Biophys. Res. Commun.*, 155:278–82.

Downey, J. M. (1990). Free radicals and their involvement during long-term myocardial ischemia and reperfusion. *Ann. Rev. Physiol.*, 52:487–504.

Entman, M. L., Youker, K., Shappell, S. B., Siegel, C., Rothlein, R., Dreyer, W. J., et al. (1990). Neutrophil adherence to isolated adult cardiac myocytes. Evidence for a CD18-dependent mechanism. *J. Clin. Invest.*, 85:1497–1506.

Friedl, H. P., Till, G. O., Tretz, O., and Ward, P. A. (1989a). Roles of histamine, complement and xanthine oxidase in thermal injury of skin. *Am. J. Pathol.*, 135:203–17.

Friedl, H. P., Till, G. O., Ryan, U. S., and Ward, P. A. (1989b). Mediator-induced activation of xanthine oxidase in endothelial cells. *FASEB J.*, 32512–18.

Granger, D. N. (1988). Role of xanthine oxidase and granulocytes in ischemia-reperfusion injury. *Am. J. Physiol.*, 255:H1269–75.

Granger, D. N., Rutili, G., and McCord, J. M. (1981). Superoxide radicals in feline intestinal ischemia. *Gastroenterology*, 81:22–9.

Granger, D. N., Benoit, J. N., Suzuki, M., and Grisham, M. B. (1989). Leukocyte adherence to venular endothelium during ischemia-reperfusion. *Am. J. Physiol.*, 257:G704–08.

Grisham, M. B., and Granger, D. N. (1989). Metabolic sources of reactive oxygen metabolites during oxidant stress and ischemia and reperfusion. *Clin. Chest Med.*, 10:71–81.

Grisham, M. B., Hernandez, L. A., and Granger, D. N. (1989). Adenosine attenuates ischemia-reperfusion-induced leukocyte adherence and extravasation. *Am. J. Physiol.*, 257:H1334–39.

Halliwell, B., and Gutteridge, J. M. C. (1989). *Free Radicals in Biology and Medicine*. Oxford: Clarendon Press.

Hamburger, S. A., and McEver, R. P. (1990). GMP-140 mediates adhesion of stimulated platelets to neutrophils. *Blood*, 75:550–54.

Hearse, D. J. Reperfusion of ischemic myocardium. (1977). *J. Mol. Cell. Cardiol.*, 9:605–15.

Huizer, T. (1989). Urate production in the heart. *J. Mol. Cell. Cardiol.*, 21:691–95.

Inauen, W., Suzuki, M., and Granger, D. N. (1989). Mechanisms of cellular injury: Potential sources of oxygen free radicals in ischemia/reperfusion. *Microcirc. Endothelium Lymphatics*, 5:143–55.

Inauen, W., Granger, D. N., Meininger, C. J., Schelling, M. E., Granger, H. J., and Kvietys, P. R. (1990). Anoxia-reoxygenation-induced, neutrophil-mediated endothelial cell injury: role of elastase. *Am. J. Physiol.*, 259:H925–31.

Jutila, M. A., Berg, E. L., Kishimoto, T. K., Picker, L. J., Bargatze, R. F., Bishop, D. K., et al. (1989). Inflammation-induced endothelial cell adhesion to lymphocytes, neutrophils, and monocytes. Role of homing receptors and other adhesion molecules. *Transplantation*, 48:727–31.

Kehrer, J. P., Piper, H. M., and Sies, H. (1987). Xanthine oxidase is not responsible for reoxygenation injury in isolated-perfused rat heart. *Free Radic. Res. Commun.*, 3:69–78.

Kishimoto, T. K., Larson, R. S., Corbi, A. L., Dustin, M. L., Staunton, D. E., and Springer, T. A. (1990). The leukocyte integrins. *Adv. Immunol.*, 46:149–82.

Korthuis, R. J., and Granger, D. N. (1986). Ischemia-reperfusion injury: Role of oxygen-derived free radicals. In A. E. Taylor, S. Matalon, and P. A. Ward, eds., *Physiology of Oxygen Radicals*. Bethesda: American Physiological Society, pp. 217–50.

Korthuis, R. J., Granger, D. N., Townsley, M. I., and Taylor, A. E. (1985). The role of oxygen-derived free radicals in ischemia-induced increases in canine skeletal muscle vascular permeability. *Circ. Res.*, 57:599–609.

Korthuis, R. J., Smith, J. K., and Carden, D. L. (1989). Hypoxic reperfusion attenuates postischemic microvascular injury. *Am. J. Physiol.*, 256:H315–19.

Korthuis, R. J., Kubes, P., Tso, P., Perry, P., and Granger, D. N. Transport kinetics for superoxide dismutase and catalase between plasma and interstitial fluid in the rat small intestine.

Kubes, P., Ibbotson, G., Russell, J., Wallace, J. L., and Granger, D. N. (1990). Role of platelet-activating factor in ischemia/reperfusion-induced leukocyte adherence. *Am. J. Physiol.*, 259:G300–05.

Lewis, M. S., Whatley, M. S., Cain, P., McIntyre, T. M., Prescott, S. M., and Zimmerman, G. M. (1988). Hydrogen peroxide stimulates the synthesis of platelet-activating factor by endothelium and induces cell-dependent neutrophil adhesion. *J. Clin. Invest.*, 82:2045–55.

Lindsay, T., Walker, P., Mickle, D., and Romaschin, A. (1988). Measurement of hydroxy conjugated dienes after ischemia/reperfusion in canine skeletal muscle. *Am. J. Physiol.*, 254:H578–83.

Lucchesi, B. R. (1990). Modulation of leukocyte-mediated myocardial reperfusion injury. *Ann. Rev. Physiol.*, 52:561–76.

Mason, R. P., and Morehouse, K. M. (1989). Electron spin resonance investigations of oxygen-centered free radicals in biological systems. In M. G. Simic, K. A. Taylor, J. F. Ward and C. von Sonntag, eds. *Oxygen Radicals in Biology and Medicine*. vol 49. New York: Plenum Press, pp. 75–79.

MacLouf, J. A., and Murphy, R. C. (1988). Transcellular metabolism of neutrophil-derived leukotriene A_4 by human platelets. A potential cellular source of leukotriene C_4. *J. Biol. Chem.*, 263:174.

McCord, J. M. (1988). Free radicals and myocardial ischemia: Overview and outlook. *Free Radic. Biol. Med.*, 4:9–14.

McCord, J. M., Roy, R. S., and Schaeffer, S. W. (1985). Free radicals and myocardial ischemia: The role of xanthine oxidase. *Advances in Myocardiology*, 5:183–89.

McKelvey, T. G., Hollwarth, M. E., Granger, D. N., Engerson, T. D., Landler, U., and Jones, H. D. (1988). Mechanisms of conversion of xanthine dehydrogenase to xanthine oxidase in ischemic rat liver and kidney. *Am. J. Physiol.*, 254:G753–60.

Moorhouse, P. C., Grootveld, M., Halliwell, B., Quinlan, J. G., and Gutteridge, J. M. (1987). Allopurinol and oxypurinol are hydroxyl radical scavengers. *FEBS Lett.*, 213:23–28.

Morris, J. B., Haglund, U., and Bulkley, G. B. (1987). The protection from postischemic injury by xanthine oxidase inhibition: blockade of free radical generation or purine salvage. *Gastroenterology*, 92:1542–47.

Parks, D. A., and Granger, D. N. (1986a). Contributions of ischemia and reperfusion to mucosal lesion formation. *Am. J. Physiol.*, 250:G749–53.

Parks, D. A., and Granger, D. N. (1986b). Xanthine oxidase: Biochemistry, distribution, and physiology. *Acta Physiol. Scand.*, 548:87–100.

Phan, S., Gannon, D., Varani, J., Ryan, U., and Ward, P.A. (1989). Xanthine oxidase activity in rat pulmonary artery endothelial cells and its alteration by activated neutrophils. *Am. J. Pathol.*, 134:1201–11.

Phillips, M. L., Nudelman, E., Gaeta, F. C. A., Perez, M., Singhal, A. K., Hakomori, S.-I. (1990). ELAM-1 mediates cell adhesion by recognition of a carbohydrate ligand, sialyl-Le^x. *Science*, 250:1130–32.

Reimer, K. A., Murray, C. E., and Richard, V. J. (1989). The role of neutrophils and free radicals in the ischemic-reperfused heart: Why the confusion and controversy? *J. Mol. Cell. Cardiol.*, 21:1225–39.

Schmid-Schonbein, G. W. (1987). Capillary plugging by granulocytes and the no-reflow phenomenon in the microcirculation. *Fed. Proc.*, 46:2397–401.

Schmid-Schonbein, G. W., Usami, S., Skalak, R., and Chien, S. (1980). The interaction of leukocytes and erythrocytes in capillary and postcapillary vessels. *Microvasc. Res.*, 19:45–70.

Seewaldt-Becker, E., Rothlein, R., and Dammgen, J. (1989). CDw18 dependent adhesion of leukocytes to endothelium and its relevance to cardiac reperfusion. In T. A. Springer, D. C. Anderson, A. S. Rosenthal, and R. Rothlein, eds. *Leukocyte Adhesion Molecules: Structure, Function, Regulation*. New York: Springer-Verlag, pp. 138–48.

Sekizuka, E., Benoit, J. N., Grisham, M. B., and Granger, D. N. (1989). Dimethylsulfoxide prevents chemoattractant-induced leukocyte adherence. *Am. J. Physiol.*, 256:H594–97.

Smith, J. K., Carden, D. L., and Korthuis, R. J. (1989). Role of xanthine oxidase in postischemic microvascular injury in skeletal muscle. *Am. J. Physiol.*, 257:H1782–89.

Smith, J. K., Carden, D. L., and Korthuis, R. J. (1991). Activated neutrophils increase microvascular permeability in skeletal muscle: role of xanthine oxidase. *J. Appl. Physiol.*, 70:2003–09.

Soussi, B., Idstrom, J.-P., Schersten, T., and Bylund-Fellenius, A.-C. (1990). Cytochrome c oxidase and cardiolipin alterations in response to skeletal muscle ischemia and reperfusion. *Acta Physiol. Scand.*, 138:107–14.

Springer, T. A. (1990). Adhesion receptors of the immune system. *Nature*, 346:425–34.

Springer, T. A., Anderson, D. C., Rosenthal, A. S., and Rothlein, R. (1989). *Leukocyte Adhesion Molecules: Structure, Function, Regulation*. New York: Springer-Verlag.

Suematsu, M., Kurose, I., Asako, H., Miura, S., and Tsuchiya, M. (1989). *In vivo* visualization of oxyradical dependent photoemission during endothelium-granulocyte interaction in microvascular bed treated with platelet-activating factor. *J. Biochem.*, 106:355–60.

Suzuki, M., Inauen, W., Kvictys, P. R., Grisham, M. D., Meininger, C., Schelling, M. E., Granger, H. J., and Granger, D. N. (1989). Superoxide mediates reperfusion-induced leukocyte-endothelial cell interactions. *Am. J. Physiol.*, 257:H1740–45.

Turrens, J. F., Crapo, J. D., and Freeman, B. A. (1984). Protection against oxygen toxicity by intravenous injection of liposome-entrapped catalase and superoxide dismutase. *J. Clin. Invest.*, 73:87–95.

Vedder, N. B., Winn, R. K., Rice, C. L., Chi, E. Y., Arfors, K.-E., and Harlan,

J. M. (1990). Inhibition of leukocyte adherence by anti-CD18 monoclonal antibody attenuates reperfusion injury in the rabbit ear. *Proc. Natl. Acad. Sci.*, 87:2643–46.

Ward, P. A., Cunningham, T. W., McCullough, K. K., Phan, S. H., Powell, J., and Johnson, K. J. (1988). Platelet enhancement of O_2 responses in stimulated human neutrophils. Identification of platelet factor as adenine nucleotide. *Lab. Invest.*, 58:37–47.

Weiss, S. J. (1989). Tissue destruction by neutrophils. *N. Engl. J. Med.*, 320:365–76.

Winterbourne, C. C. (1986). Myeloperoxidase as an effective inhibitor of hydroxyl radical production. Implications for the oxidative reactions of neutrophils. *J. Clin. Invest.*, 78:545–50.

Zimmerman, B. J., Grisham, M. B., and Granger, D. N. (1990). Role of leukotriene B_4 in ischemia/reperfusion-induced granulocyte infiltration. Gastroenterology, (in press).

4

The Oxidation of Lipoproteins: Implications for Atherosclerosis

GUY M. CHISOLM III

LIPOPROTEINS AND ATHEROSCLEROSIS: THE MISSING LINK

Low-Density Lipoproteins, High-Density Lipoproteins, and Atherosclerosis

Most basic researchers and physicians who are involved with the study of atherosclerosis or its treatment no longer doubt that plasma cholesterol concentration is a potent risk factor for coronary disease. The ways in which cholesterol and its lipoprotein carriers act to transform a healthy segment of artery wall into an atherosclerotic lesion that can ultimately occlude blood flow in the heart have been intensely studied for decades. Although much has been learned about the interactions between lipoproteins and arterial tissue, the molecular and cellular mechanisms in the disease process are still poorly understood. Low-density lipoprotein (LDL) is typically the major cholesterol-carrying lipoprotein in human plasma, and it is the concentration of this lipoprotein that correlates best with the risk of heart disease. Despite intriguing associations between LDL and atherosclerosis, LDL does not interact with vascular cells in ways that would be anticipated if LDL actually caused atherosclerosis. Thus, on a molecular and cellular level, the link between a high plasma LDL level and atherogenesis has remained obscure.

In the last decade, data have accumulated in support of a theoretical candidate for the missing link. These data implicate a chemical alteration of LDL by the action of reactive oxygen metabolites (Jurgens et al. 1987; Steinberg et al. 1989; Steinbrecher et al. 1990). In this theory, it is the oxidatively modified form of LDL that facilitates the vascular disease process. Unlike native (or unmodified) LDL, oxidized LDL interacts with cells in ways that are consistent with features that have long been known to be typical of atherosclerotic lesions. In the present paper, after providing a brief background of lipoproteins and atherosclerosis we present a lipoprotein oxidation theory of atherosclerosis. We then discuss the effects of oxidized LDL on cells, the evidence that oxidized LDL exists in vivo, and the evidence that its existence has pathologic consequences.

Human Lipoprotein Classification

Human lipoproteins are typically categorized as chylomicrons, very low density lipoprotein (VLDL), LDL, and high-density lipoprotein (HDL). These populations of lipoprotein particles are generally separated from plasma by ultracentrifugation according to their hydrated densities. In addition, a population of particles intermediate in density between VLDL and LDL, called intermediate density lipoprotein, is often given a separate category. Lipoprotein composition can be defined in broad categories, in that each of the lipoprotein groups has particular protein moieties as well as different amounts of triglyceride, phospholipid, cholesteryl ester, and cholesterol. The lipoproteins vary markedly in their respective proportions of these lipids; chylomicrons and VLDL are triglyceride rich, whereas HDL has minimal triglyceride and a relatively high proportion of protein and phospholipid. Cholesteryl esters and triglycerides, the least polar of the lipids, reside in the hydrophobic core of the lipoproteins, whereas phospholipids with their more polar head groups are oriented at the surface. The lipoproteins also vary with respect to the apolipoprotein moiety. Chylomicrons, VLDL, and LDL all contain apolipoprotein B. Chylomicrons and VLDL have other apolipoproteins in addition to apoB, whereas the apolipoprotein moiety of LDL is almost exclusively apoB. The major apolipoproteins of HDL are apolipoprotein A-1 and A-2; minor ones (in quantity, not in metabolic importance) are apolipoproteins C-I, C-II, and C-III. Apolipoprotein E is an important lipoprotein found in various lipoprotein families.

It is misleading to regard the lipoprotein categories described above as homogeneous. Each category has significant heterogeneity in size and composition. For example, the lipid classes can vary in composition. Cholesteryl esters, phospholipids, and triglycerides all contain variations in their fatty acid compositions, which are, to a certain extent, influenced by diet. Phospholipids vary additionally in their head groups; LDL, for example, has variable amounts of phosphatidyl choline, sphingomyelin, phosphatidylethanolamine, phosphatidylserine, and phosphatidyl inositol. Subclasses of HDL exist with different apolipoproteins predominating; there are, for example, HDL_1 and HDL_2, which are separable by their hydrated densities and which differ in the proportions of HDL apolipoproteins present and vary in their metabolic activities.

Metabolism contributes further to the heterogeneity of the lipoprotein groups. Chylomicrons are assembled in gut epithelium after absorption of cholesterol and fatty acids from the diet. The epithelial cells produce a form of apolipoprotein B called apoB-48. It has a lower molecular weight than the apolipoprotein called apoB-100, which is produced by hepatocytes and is incorporated into VLDL. Chylomicrons first pass through the mesenteric lymph and then enter the blood at the vena cava by means of the thoracic duct. Chylomicrons and VLDL in circulating blood are more appropriately referred to as chylomicron remnants or VLDL remnants, because endothelial cell-bound lipoprotein lipase hydrolizes the triglyceride moiety for storage in adipocytes. In addition, the plasma-borne enzyme, lecithin: cholesterol acyl transferase, esterifies the free (unesterified) cholesterol of HDL, and lipid transfer protein facilitates the transfer of cholesteryl ester and triglyceride among the various lipoproteins. Those lipids that are slightly less hydrophobic, the phospholipids and cholesterol, exchange more freely, not only among lipoproteins, but also between lipoproteins and cell membranes.

The hydrolysis by lipoprotein lipase of triglyceride in VLDL leaves LDL as a circulating metabolic end product. It is the plasma level of this heterogenous lipoprotein class that correlates well with the risk of atherosclerosis. High levels of HDL, on the other hand, are believed to be protective against vascular disease because of their role in "reverse cholesterol transport," the removal of cholesterol from sites in peripheral tissues and transport of this cholesterol back to the liver.

Lipoproteins in Experimental Atherosclerosis

It is important to note parenthetically that feeding experimental animals diets enriched in cholesterol does, in many species, result in the formation of early, foam-cell rich fatty streak lesions in the animals' arteries. The lipoprotein elevated in these animals, however, is not solely LDL. An aberrant form of VLDL, called beta-VLDL (β-VLDL, named because of its β-migration upon electrophoresis, which is distinct from the pre-β-migration of normal VLDL), is produced in large amounts and is markedly elevated, even in the plasma of fasting subjects. β-VLDL is apolipoprotein E-rich, is taken up readily by macrophages, and is therefore believed to be atherogenic.

Thus, although much information continues to be learned about atherosclerosis from studying diet-induced lesions in animals, questions relating to the mechanisms by which LDL participates in lesion development in vivo may be better answered by studying experimental models such as the Wanatabe heritable hyperlipemic (WHHL) rabbit, in which an LDL receptor defect leads to high LDL levels and spontaneous atherosclerosis.

Features of the Atherosclerotic Lesion

An artery wall is divided into three anatomically distinct, concentric regions: the intima, media, and adventitia. Arteries, like all of the blood vessels of the body, are lined with a monolayer of endothelial cells. The endothelial cell has capabilities for a broad array of functions that regulate blood tissue interactions, including the exchange of molecules between tissue and blood and the responses to injury. The endothelium is attached to a collagenous basement membrane, underlying which is interstitial space containing proteoglycans and collagen and occasionally smooth muscle cells. This region is known as the intima. In normal vessels, the intima is very thin, so thin, in fact, that the endothelium appears to be immediately adjacent to the elastin structure separating the intima from the media—the internal elastic lamina. In both muscular and elastic arteries, concentric layers of vascular smooth muscle cells make up the next anatomic portion of the artery wall—the media. The media ends and the adventitia begins at another elastic layer—the external elastic lamina. In elastic arteries, an additional, variable number of elastin layers separate the layers of smooth muscle cells. The adventitia is made up of connective tissue, fibroblasts, and adipocytes. The adventitia and, in large arteries, the outer media are vascularized with a capillary network called the vasa vasorum.

Atherosclerosis occurs in both muscular (for example, the coronary arteries) and elastic (for example, the aorta) arteries; in early stages the disease grossly affects only the intima. A general, though not universally held categorization divides atherosclerotic lesions into fatty streaks, fibrous lesions, and advanced or complicated lesions. These designations apply to human lesions, but have counterparts in animal models. Because of the comparatively short duration of animal experiments and because of the heavier emphasis in research on probing atherogenesis instead of lesion development, the earlier fatty streak lesion is most often studied in animal models. In susceptible animals, the fatty streak appears in the aortic intima after a short period of cholesterol supplementation to the diet. There is an intact endothelium over the fatty streak, although there is some evidence that the endothelium has a higher turnover (that is, occurrence of death and compensatory cell division) at the sites where fatty streaks usually occur than at less disease-prone sites (Caplan and Schwartz 1973). The fatty streak has large numbers of monocyte-derived macrophages in the markedly thickened intima (Gerrity 1981; Ross 1986). These so-called foam cells are engorged with numerous lipid droplets in which cholesteryl esters predominate. The interstitial space of the intima contains a relatively high concentration of LDL (Hoff and Gaubatz 1982; Smith and Ashall 1983), and the source of the macrophage lipid is known to be the plasma lipoproteins.

The fibrous lesion consists of a core made up of lipid-laden foam cells of either macrophage or smooth muscle cell origin, dead cell debris, and extracellular lipid deposition that can include crystalline cholesterol. This lesion is also characterized by a proliferative component between the intact endothelium and the core of the lesion. This proliferative component consists of vascular smooth muscle cells that have migrated from the media and proliferated vigorously (Ross and Glomset 1973). The incidence of smooth muscle cell death is thought to be elevated as well, but the kinetics of cell replication and cell death are such that a large number of smooth-muscle-derived cells accumulate in the "fibrous cap" of this lesion (Thomas et al. 1968). The interstitial matrix of this fibrous lesion is vastly expanded, and proteoglycans and collagen are produced at high rates and accumulate, giving the lesion its fibrous characteristics.

The stages of lesion development are not universally agreed on. In general, it is believed that the fatty streak can, but does not necessarily, develop further into a fibrous lesion. Reversibility of atherosclerosis has been documented both in humans and in animal models of atherosclerosis, but the process is not completely understood. In general, it is believed that the fatty streak is a reversible lesion under appropriate circumstances whereas the fibrous plaque can decrease in size, but does not disappear. The complicated lesion can be ulcerated with platelet deposition at sites of physically disrupted or sloughed endothelium (Ross 1986). Late-stage lesions are often found to contain deposits of insoluble, oxidized lipid- and protein-containing extracellular material called lipofuscin or ceroid.

The formation of atherosclerotic lesions is believed to involve an injury response in endothelium (Ross 1986), characterized by monocyte adherence (Gerrity 1981) and altered regulation of lipoprotein exchange between blood and intima. A growth factor effect involving migration and proliferation of smooth muscle cells also occurs, as does, in late stages, angiogenesis. There is, in addition, a lipoprotein-induced effect in which metabolism of large quantities of lipoprotein-derived lipids takes place, resulting in cholesterol esterification and accumulation in macrophages and smooth muscle cells and the transformation of these cells into foam cells.

The Potential Role of LDL Oxidation in Atherosclerosis: A Hypothesis

Although it is known that lipoproteins supply the substrate for the lipid deposition of early atherosclerosis, the mechanism by which LDL brings about the arterial changes characteristic of the fatty streak are not known. The obscurity is due, in part, to the fact that in experiments using cultured cells, plasma LDL does not cause monocytes to adhere to endothelium; it is not a chemoattractant for monocytes; it does not injure endothelial cells; it does not induce smooth muscle cell proliferation; and it does not lead to foam cell formation in either macrophages or smooth muscle cells.

In remarkable contrast, the changes brought about in the way LDL interacts with cells after the oxidative modification of the lipoprotein can be used to construct a theory of the role of oxidized LDL in atherogenesis. Although the cellular interactions of oxidized LDL fit better with the known features of atherosclerosis development than do those of unaltered LDL, it is important to emphasize the speculative nature of the hypothesis. As we will show later, the evidence that oxidized LDL exists in atherosclerotic lesions is compelling; however, the data currently available that support a causal role for oxidized LDL in vascular disease are only suggestive. It is also important to indicate that the hypothesis presented below is but one of a number of possible variations of theories that link LDL oxidation to atherosclerosis, and it is offered as an example.

In the version of the hypothesis presented here, the initial predisposing condition for atherosclerosis is an elevated level of plasma LDL. It was demonstrated previously that elevated plasma LDL levels result in elevated interstitial LDL levels in arterial tissue. An example of this phenomenon was reported in experiments in which the ratio of LDL in rabbit aortic intima and media to LDL in plasma was the same before and after the injection of sufficient LDL to raise the plasma cholesterol two- to threefold (Bratzler et al. 1977). It has been predicted mathematically that the residence time an LDL molecule would spend in arterial interstitial space is increased significantly if the plasma (and therefore, tissue) LDL level is elevated (Schwenke and Carew 1989). Thus, a consequence of elevated interstitial LDL in this scenario is an increase in its residence time in intercellular spaces and thereby an increased probability of its exposure to oxidative modification. The rationale for proposing that an oxidative modification to LDL in the arterial intima might occur is the demonstration in vitro that vascular smooth muscle cells (Morel et al. 1984; Heinecke et al. 1984) and certain endothelial cells (Morel et al. 1984; Steinbrecher et al. 1984) are capable of oxidizing LDL readily, particularly in the relative absence of other plasma proteins and lipoproteins. It is known that, early in the development of atherosclerosis, elevated levels of interstitial LDL relative to other plasma proteins exist in arteries, in proximity to these two cell types (Schwenke and Carew 1989).

A concomitant event in the early development of experimental atherosclerosis is the adherence of blood-borne monocytes to vascular endothelium, their diapedesis through the endothelium, and their accumulation within the intima in large numbers in focal areas (Gerrity 1981; Ross 1986). That oxidized LDL may play a role in the recruitment of monocytes is suggested by data that show that moderately oxidized LDL, but not unaltered plasma LDL, induces monocyte adherence to endothelial cells in culture (Berliner et al. 1990) and induces the production by both endothelial cells and smooth muscle cells of monocyte chemotactic protein-1 (Cushing et al. 1990). In addition, severely oxidized LDL itself is a chemoattractant for human monocytes (Quinn et al. 1985, 1988) but an inhibitor of migration for macrophages in culture (Quinn et al. 1987). This latter result implies that oxidized LDL or its components may play a role in the immobilization of macrophages in the intima and, by this action, support the expansion of the lesion as large numbers of monocyte-derived macrophages occupy the intimal spaces. It has been demonstrated in cell culture systems that macrophages harvested from the mouse peritoneum (Parthasarathy et al. 1986a) and human monocyte-derived macrophages given a phagocytic stimulus (Cathcart et al. 1985, 1989) can also readily oxidize LDL. This finding suggests a mechanism by which the in vivo situation in early atherosclerosis may worsen. Further oxidation of LDL may be greatly facilitated by the contribution of free radicals from macrophages, and, in the context of the arterial intima, this contribution may be synergistic with the aforementioned action of smooth muscle cells and endothelium in oxidizing the high concentration of interstitial LDL known to be adjacent to these three cell types in early lesions (Hoff and Gaubatz 1982; Smith and Ashall 1983; Schwenke and Carew 1989). Severely oxidized LDL is recognized by scavenger receptors expressed on the surface of endothelial cells and macrophages (Steinbrecher et al. 1984; Freeman et al. 1991). These findings allow speculation that oxidized LDL is the in vivo ligand leading to the unregulated uptake of LDL-borne lipid by intimal macrophages and the transformation of these cells to the typical morphologic pattern of foam cells.

The participation of oxidized LDL as a cytotoxin can also be proposed to play a role in the atherosclerotic sequence (Jurgens et al. 1987; Hessler et al. 1979, 1983; Morel et al. 1983). Early in atherogenesis, oxidized LDL could be responsible for the increases in endothelial cell turnover reported to occur at lesion-prone sites of the arterial system (Caplan and Schwartz 1973). As a cell-injuring agent, oxidized LDL could thus be incorporated into the evolving theories of atherosclerosis built around endothelial cell injury (Ross, 1986). Oxidized LDL, in addition to injuring endothelial cells as an early event, may be responsible for the dead cell debris that has been described in the core of atherosclerotic lesions. Which, if any, of the tenets of the foregoing theory will be borne out by critical experimental testing is not now known. A description, derived predominantly from in vitro studies, of the known cellular interactions of LDL and oxidatively modified LDL, that indicate a wide spectrum of potential cell-modifying activities, follows.

CELL RECEPTORS FOR LIPOPROTEINS

Low-Density and High-Density Lipoprotein Receptors

Several distinct binding sites for lipoproteins have been identified that dramatically affect the way in which cells and lipoproteins interact. The first to be discovered and the most thoroughly characterized is the LDL receptor (Goldstein and Brown 1977). In addition, the acetylated LDL or scavenger receptor has been characterized (Brown and Goldstein 1983), and binding sites for HDL are currently being studied (Biesbroeck et al. 1983). Most recently, the LDL-receptor-related protein, which mediates cellular interactions for lipoproteins enriched in apolipoprotein E, has been identified (Kowal et al. 1989). All of these receptors regulate or alter cellular cholesterol metabolism.

The LDL receptor, which binds lipoproteins containing apolipoproteins B and E, is present in abundance on normal human liver cells and on peripheral cells. A cell deprived of lipoprotein, such as a cultured vascular smooth muscle cell, a fibroblast, or a monocyte incubated in lipoprotein-deficient media for several hours, will synthesize cholesterol from acetate by a cascade of enzymatic steps, and, in addition, will express relatively large numbers of LDL receptor molecules on its plasma membrane (Goldstein and Brown 1977). The addition of LDL to the medium results in lysosomal hydrolysis of the LDL including the degradation of cholesteryl esters to cholesterol and the entry of this free cholesterol into cellular pools. The added cholesterol inhibits cellular cholesterol synthesis, down-regulates the synthesis of new LDL receptors, and promotes esterification of excess cholesterol. In other words, the LDL receptor provides the animal with a means of LDL degradation by an efficient process and simultaneously provides the cell with exogenous cholesterol in lieu of its own cholesterol synthesis. LDL receptor deficiencies occurring in both heterozygous or homozygous familial hypercholesterolemia, are associated with high LDL cholesterol levels, increased risk of atherosclerosis, and premature myocardial infarction (Goldstein and Brown 1977).

In the context of the oxidative modification of LDL, it should be pointed out that LDL oxidation can result in protein alterations in which lipid peroxidation products, such as the aldehydes produced from fatty acid oxidation, can covalently bind to lysyl residues of apolipoprotein B (Jurgens et al. 1987). Lysyl modification of apolipoprotein B interferes with the LDL receptor recognition site on the apoprotein (Jessup et al. 1986). Thus, whereas mild oxidation may not interfere quantitatively with LDL binding to its receptor, further oxidation progressively inhibits receptor recognition until there is none. Further oxidation leads to marked changes in the surface charge of LDL and results in scavenger receptor recognition of the modified lipoprotein (Steinbrecher et al. 1984; Freeman et al. 1991).

The binding sites for HDL promote movement of cholesterol among intracellular pools and may thereby facilitate efflux of cholesterol from cells (Slotte et al. 1987). The alterations in the interactions of HDL with its receptor after oxidation of HDL have not been thoroughly studied. Interactions of apolipoprotein E-containing lipoprotein with the LRP may, like the interactions of LDL with the LDL receptor, mediate the cellular uptake and hydrolysis of cholesteryl esters. The amount of LRP expressed by cells does not, however, appear to be reduced by the introduction to the cell of exogenous sterol. It is not currently known what changes occur in the LRP-ligand interactions if the lipoprotein ligand is oxidatively modified.

The Scavenger Receptor

The scavenger receptor was identified on macrophages that "recognized" and degraded chemically modified forms of LDL in which the lysyl residue of the apolipoprotein B-100 moiety of LDL was covalently bound to certain chemical adducts and the resulting charge of LDL was increased (Brown and Goldstein 1983). This receptor recognizes numerous modified forms of LDL, including acetylated LDL, malondialdehyde-modified LDL, and LDL altered by free radical oxidation (Steinbrecher 1987; Brown and Goldstein 1983; Haberland et al. 1982; Freeman et al. 1991). The accumulation of cholesteryl esters in macrophages, which follows receptor recognition and uptake of these modified lipoprotein ligands, does not downregulate the expression of the receptor (Brown and Goldstein 1983). This process leads to accumulation of lipid on a level that is consistent with foam-cell formation. Scavenger receptors have been demonstrated on confluent endothelial cells (Sanan et al. 1985) as well as on macrophages.

LIPOPROTEIN OXIDATION IN A CELL-FREE SYSTEM

Lipoproteins are readily oxidized in vitro, even under cell-free conditions (Hessler et al. 1983; Morel et al. 1984). Lipoprotein isolation has traditionally been performed after the immediate addition to blood or plasma of compounds that protect the vulnerable lipids from oxidation. The cation chelator, ethylenediaminetetraacetic acid (EDTA), commonly is added either to blood, to inhibit coagulation, or to plasma, if other anticoagulants are used. The presence of sufficient EDTA throughout the isolation procedure can inhibit gross lipid peroxidation by impeding the action of metal ions that catalyze lipid peroxidation. In the absence of such protection, LDL easily oxidizes to the degree that it becomes toxic to cultured cells (Hessler et al. 1983; Morel et al. 1984). Early reports of cell injury by LDL can in many cases be attributed to the inadvertent oxidation taking place when, for example, the LDL was dialyzed free of EDTA in preparation for adding it to the medium of cell-culture experiments. The degree to which LDL oxidation occurs in various conditions in the absence of protective antioxidants, chelators, or free radical scavengers, and the abilities of vitamin E, butylated hydroxytoluene, superoxide dismutase (SOD), reduced glutathione, and other agents to block the oxidation have been reported (Heinecke et al. 1984; Steinbrecher et al. 1984; Hessler at al. 1983; Morel et al. 1984). It is clear that the presence of iron and copper ions facilitates lipoprotein oxidation (Heinecke et al. 1984; Steinbrecher et al. 1984; Kosugi et al. 1987). This fact is not surprising given the extensive literature relating the role of metal ions in the initiation of lipid peroxidation and given the variety and quantity of reactive lipids that make up both the surface and the core of lipoprotein particles.

The recognition of the role metal ions can play in LDL oxidation becomes particularly important in the interpretation of results from studies of lipoprotein-cell interactions. The commercial media commonly used to supply nutrients to cultured cells contain variable concentrations of ferrous and cupric ions, as well as a number of substances that would contribute antioxidant effects (for example, selenium). The results of an experiment examining particular effects of LDL on a specific cell type might vary if conducted in a medium supporting LDL oxidation, such as Ham's F-10, which contains relatively abundant cupric ion, rather than in a medium relatively low in metal ions.

For similar reasons, the constituency of cell culture media influences experiments designed to examine the oxidation of LDL by cells. Free radical production by cultured human vascular endothelial cells (Morel et al. 1984; Steinbrecher et al. 1984), by vascular smooth muscle cells from a number of species (Morel et al. 1984; Heinecke et al. 1984), by cultured peritoneal macrophages (Parthasarathy et al. 1986a), and by adherent populations of human monocytes offered a phagocytic stimulus (Cathcart et al. 1985) is sufficient to oxidize LDL. These findings are dependent on the level of LDL oxidation measured in the medium in which the cells are incubated being significantly higher than the "background" oxidation level of LDL measured in the same medium in a cell-free dish. This background oxidation is likely to be higher in experiments using medium containing cupric ions, and, in certain cell culture systems, the contribution of the particular cell to oxidation may actually depend on the presence of the metal ion in the medium. Although the role of the medium has been acknowledged (Parathasarathy 1987), the degree to which it contributes to particular experimental findings often is not thoroughly analyzed.

The changes taking place when LDL becomes oxidized are significant but, to date, little has been reported to indicate that LDL oxidized in a cell-free system differs functionally from LDL oxidized by any of the various cell types, other than differences attributable to the degree of oxidation (Morel et al. 1984; Steinbrecher 1987; Steinbrecher et al. 1987). Certain characteristic physiochemical changes accompanying on LDL oxidation in vitro can be predicted. Significant changes take place in the grossly apparent physical features of human LDL. The golden color typical of human LDL isolated by ultracentrifugation fades with oxidation to a pale yellow, then to translucence. Further oxidation in concentrated preparations of LDL (for example, mg LDL cholesterol/ml) leads to grossly visible aggregation. Chemical measurements reveal the loss of polyunsaturated fatty acids (Esterbauer et al. 1987), losses of free cholesterol because of the formation of its various oxidized derivatives (Steinbrecher 1987), losses in the tocopherols including vitamin E (Esterbauer et al. 1987), and a loss in total cholesterol ascribed to decreases in cholesteryl ester (Steinbrecher 1987). Loss of polyunsaturated fatty acids is accompanied by the formation of numerous fatty acid oxidation derivatives, such as malondialdehyde and 4-hydroxynonenal (Esterbauer et al. 1987; Quehenberger et al. 1987). The more polar, aqueous-soluble derivatives like malondialdehyde may enter the aqueous solvent and leave the LDL complex, although some of the aldehydes formed can react with available lysyl residues (Esterbauer et al. 1987), altering the structure of the apolipoprotein and the surface charge of the complex (Jurgens et al. 1987). Additional changes to the protein include a marked fragmentation due to nonenzymatic degradation (Fong et al. 1987). A phospholipase A_2 activity of LDL, with a substrate specificity favoring oxidized phospholipids, has been discovered through studies of

LDL oxidation (Steinbrecher et al. 1984; Parathasarathy et al. 1985; Stafforini et al. 1987; Steinbrecher and Pritchard 1989). This enzyme activity results in significant generation of lysophosphatidyl choline during LDL oxidation. The foregoing changes result collectively in increases in the hydrated density of the lipoprotein complex and in the negativity of its surface charge as shown by increased electrophoretic mobility (Henriksen et al. 1981).

In the early stages of research on the altered effects of LDL on cell function after LDL oxidation, oxidized LDL was considered an actual entity because of the lack of sufficient characterization. Recent results have revealed that the numerous effects attributable to oxidized LDL may manifest themselves at different degrees of oxidation. For example, it is known that as oxidation of LDL proceeds, the ability of LDL to be recognized by its receptor decreases; however, moderately oxidized LDL with minimal change in charge can be a potent cytotoxin (Morel et al. 1983). Before the decrease in LDL receptor recognition and before the enhanced recognition by scavenger receptors, oxidized LDL can stimulate gene expression for colony stimulation factors (CSFs) in endothelial cells (Rajavashisth et al. 1990). On further oxidation, the capacity to evoke CSF gene expression is diminished. This has led the authors of the preceeding study to define "minimally-modified LDL" as LDL that is oxidized to a limited extent and that is still recognized by the LDL receptor, but not by scavenger receptors.

Presented below are some of the recently reported effects of oxidized LDL that are distinct from those for native LDL. Caution in overinterpretation is warranted because the culture medium incubating the cells varies and the degree of oxidation of LDL is by no means standardized or, in some cases, even easily discerned. Recognition of the heterogeneity of oxidized LDL is essential in attempting to interpret its effects. Native LDL contains numerous lipid classes, and in each class numerous variations exist. For example, there are several different phospholipids present on LDL, each with possible variations in fatty acid constituency. On oxidation, these dozens of compounds can become hundreds, and it is likely that the cellular changes induced by oxidized LDL may ultimately be ascribed to different molecules that are among these new lipid derivatives.

THE INTERACTIONS OF OXIDIZED LIPOPROTEINS WITH CELLS

A Role for the Scavenger Receptor

In the early reports that endothelial cells modified LDL, it was demonstrated that the modified LDL competed successfully for the degradation of acetylated LDL by murine macrophages and that acetylated LDL competed for uptake of the so-called "endothelial cell-modified LDL" (Henriksen et al. 1981). Later reports showed that EC-modified LDL was in fact LDL modified by free-radical oxidation (Morel et al. 1984; Steinbrecher et al. 1984), which led to speculation that oxidized LDL might be the in vivo ligand for the acetylated LDL receptor (Jurgens et al. 1987; Steinberg et al. 1989; Steinbrecher et al. 1990; Freeman et al. 1991). This approach implied that the scavenger receptors on macrophages exist to remove oxidatively damaged lipoproteins and other altered proteins as well. The data presented in these studies revealed that oxidized LDL and acetylated LDL did not compete with each other *completely* for uptake by macrophages (Henriksen et al. 1981). Evidence has been presented for the existence on macrophages of more than one receptor for modified LDL (Arai et al. 1989; Sparrow et al. 1989). In one study, two receptors were proposed, one which recognized both acetylated and oxidized LDL and one which recognized oxidized but not acetylated LDL (Sparrow et al. 1989). In another, three receptors were proposed, one each of which recognized either oxidized LDL or acetylated LDL exclusively, and a third which recognized both (Arai et al. 1989). More recently, two variations of the scavenger receptor have been cloned and transfected into Chinese hamster ovary cells (Kodama et al. 1990; Rohrer et al. 1990). Both of these appear to recognize acetylated LDL, indicating that a cysteine-rich region occurring at the C-terminal end of one but not the other, is not the peptide region containing the acetylated LDL binding site. An editorial accompanying these reports contained the comment that whether either or both of these receptors recognize oxidized LDL was a "crucial question" awaiting further study (Brown and Goldstein 1990). Recently, it has been demonstrated that both receptors mediate high affinity, saturable endocytosis of oxidized LDL as well as acetylated LDL (Freeman et al. 1991). Interestingly, in transfected cells, acetylated LDL efficiently competed for both its own endocytosis and that of oxidized LDL. Oxidized LDL, however, competed effectively for its own endocytosis, but only incompletely competed for endocytosis of acetylated LDL. Such nonreciprocal cross-competition may indicate separate but interacting binding sites for the two ligands on the same receptor (Freeman et al. 1991).

The existence of scavenger receptors implies that animals and humans have a capacity to recognize and remove proteins modified, or "damaged," by oxidation, a capacity that could be construed as part of the protective role of macrophages in removing unwanted or unneeded byproducts. The recognition and removal of LDL by macrophages leads to the intracellular accumulation of lipid, especially cholesterol, which, in part, becomes esterified and is stored. The uptake of oxidized LDL could thus be perceived as removal by macrophages of material that is toxic. Faced with excess material to remove, the overburdened macrophage may become a foam cell and contribute to the development of pathologic lesions in the arterial intima, and its otherwise protective capacity is thereby subverted.

Oxidized Low-Density Lipoprotein and Macrophages

Macrophage Oxidation of Low-Density Lipoprotein

That macrophages can oxidize LDL was demonstrated using human LDL exposed to adherent human monocytes in culture (Cathcart et al. 1985). In this system, a macrophage stimulus was required to induce the cells to modify LDL. The successful stimuli included endotoxin, and the phagocytic stimulus, opsonized zymosan (Cathcart et al. 1988). The general free-radical scavengers and lipophilic antioxidants, vitamin E and butylated hydroxytoluene, effectively reduced the oxidation by these stimulated monocyte-derived macrophages, implicating oxidative free radicals (Cathcart et al. 1985). Subsequent studies directed toward the mechanism of this oxidation process revealed that superoxide dismutase was effectively able to block the oxidation of LDL, but that the oxidation did not require exposure of LDL to the cells during the enhanced superoxide output taking place during the respiratory burst evoked by the stimulation (Cathcart et al. 1989; Hiramatsu et al. 1987). These findings are consistent with the fact that the lower level of superoxide emanating from the stimulated cells following the respiratory burst is sufficient to oxidize LDL (Cathcart et al. 1989). In addition, macrophage oxidation appears to depend on cellular lipoxygenase(s) inasmuch as eicosatetraynoic acid (ETYA), an arachidonic acid antagonist, blocked the macrophage oxidation of LDL, whereas cyclooxygenase inhibitors did not (McNally et al. 1990). Furthermore, those lipoxygenase inhibitors that did not appear to be general antioxidants blocked both lipoprotein oxidation by macrophages and that by soybean lipoxygenase in a cell-free system (McNally et al. 1990; Sparrow et al. 1988). From these results, the relationship between lipoxygenase activity, superoxide anion release, and LDL oxidation cannot be determined, other than to say that they are interrelated. Furthermore, it is unknown whether lipoxygenase-mediated oxidation of LDL derives from the direct action of the enzyme on LDL lipid substrates (Sparrow et al. 1988; Cathcart et al. 1991) or whether lipoxygenase action on cell-derived lipids subsequently transferred to LDL accounts for the observed lipoprotein oxidation. It should be pointed out that the experiments on macrophage oxidation of LDL discussed previously were performed largely in culture medium relatively low in metal ions. Thus, if metal ions are involved in this oxidation, they may be supplied by cellular stores that are released on stimulation.

Cultured macrophages other than the monocyte-derived populations of cells already discussed also oxidize LDL. Mouse peritoneal macrophages, for example, oxidize LDL to the extent that the LDL is taken up by the macrophages by way of a scavenger receptor-mediated process (Parthasarathy et al. 1986a). These experiments were performed in a relatively metal-ion rich medium, Ham's F-10 medium. The dependence of LDL oxidation on the medium is illustrated by a recent study in which cells grown in a metal ion-poor medium were switched to Ham's when LDL was to be oxidized; if the Ham's medium was not freshly made, ferrous ion was added to promote oxidation (DeWhalley et al. 1990).

Despite the differences among protocols and the residual ambiguities related to the culture media, it is clear that macrophages have the capacity to oxidize LDL and thus may do so in vivo. The macrophage in the early arterial foam-cell lesion—the fatty streak—is surrounded by interstitial lipoprotein (Hoff and Gaubatz 1982; Smith and Ashall 1983), leading one to conclude that the conditions prevailing are at least conducive to macrophage modification of LDL.

Oxidized Low-Density Lipoprotein Metabolism by Macrophages

There are four distinct modes by which oxidized LDL could interact with macrophages that would lead to lipoprotein internalization. Macrophages typically express the LDL receptor, which is capable of regulated uptake of moderately oxidized LDL; the scavenger receptor(s), which is capable of recognizing oxidizing LDL; and the Fc receptor, which could theoretically recognize complexes of oxidized LDL and the antibody described in human plasma that can recognize oxidized LDL (Palinski et al. 1989). Macrophages are also capable of phagocytosis if they are given an appropriate particulate stimulus. Thus, native LDL or LDL oxidized to a limited extent such that its recognition by the LDL receptor is not completely diminished, can be internalized by the LDL receptor. Severely oxidized LDL can be internalized through the scavenger receptor pathway, and LDL that is oxidized to the extent that it aggregates can be taken up by phagocytosis. The results of internalization via these pathways vary. Uptake by the LDL receptor leads to downregulation of the LDL receptor expression, decreasing the quantitative importance of this route. In contrast, internalization of modified lipoprotein via the scavenger receptor is not downregulated by the presence of high concentrations of the ligand (Brown and Goldstein 1983). This finding is consistent with the theory that oxidized LDL, but not native unmodified LDL, might be an in vivo ligand for the scavenger receptor, which, when present in high concentrations, leads to esterification and storage of lipoprotein-derived cholesterol.

There is indirect evidence that LDL may exist in atherosclerotic lesions in aggregated form, which is consistent with the idea that LDL in the intimal interstitium is being oxidized. LDL-like (apolipoprotein B-containing) material extracted from human and animal atherosclerotic lesions has been found in a monomeric form, reminiscent of the plasma lipoprotein in its spheroid appearance under the electron microscope, as well as in an aggregated or coalesced form, insoluble and more difficult to extract from the lesion material (Goldstein et al. 1981; Hoff et al. 1978). Although it cannot be completely ruled out that the latter amorphous form is an artifact of the extraction procedure, it at least allows the speculation that LDL oxidation occurs in a lesion to the point of aggregation and that the resulting aggregates may be taken up by phagocytosis and contribute to foam-cell formation.

Pathologists have been aware for decades that advanced atherosclerotic lesions often contain material (referred to as ceroid or lipofucsins) that has both proteinaceous and oxidized lipid components. It is insoluble and exhibits characteristic histologic staining properties and fluorescence. This long-standing observation, coupled with more recent evidence of oxidized lipid accumulation in vascular lesions, is again consistent with a hypothesis of atherosclerosis involving lipoprotein oxidation and aggregation (Steinberg et al. 1989).

Leukocyte Motility and Oxidized Low-Density Lipoprotein

There are several in vitro observations that relate to the effects of oxidized LDL on the attraction and immobilization of monocytes. This direction of study is being pursued vigorously because of the intimate involvement of the monocyte in the early fatty streak lesion of atherosclerosis. Oxidized LDL has been shown to be a chemoattractant for human monocytes, and the chemoattractive component has been shown to reside in the lipid portion of the lipoprotein (Quinn et al. 1985, 1988). Further experimentation has shown that lysophosphatidyl choline is a monocyte chemoattractant and may be the—or one of the—chemoattractant constituent of oxidized LDL because of the phospholipase A_2 activity known to be exhibited by LDL during oxidation (Steinbrecher et al. 1987; Quinn et al. 1988). Mouse peritoneal macrophages are immobilized by oxidized LDL (Quinn et al. 1987). If this finding also applies to human macrophages that reside in a fatty streak lesion, it suggests that oxidized LDL may be responsible for focal immobilization of foam cells that might otherwise have cleared from the vascular tissue. Oxidized LDL causes endothelial cells to produce a chemoattractant specifically for monocytes (Berliner et al. 1990; Cushing et al. 1990). A component of oxidized LDL may thus further mediate the attraction of monocytes to early lesion sites through this monocyte chemotactic peptide. Interestingly, this effect is seen with so-called "minimally modified LDL," which is, by the definition used in the studies, still recognizable by the LDL receptor (Berliner et al. 1990; Cushing et al. 1990).

Taken together, the foregoing in vitro findings support the view that the accumulation of oxidized LDL or certain of the components of oxidized LDL may control the attraction of monocytes, the accumulation of lipid from interstitial oxidized LDL, and the immobilization of lipid-laden foam cells. One might speculate that extended residence time and extensive lipid engorgement of these cells may result in eventual cell death locally, exacerbating the problem of the removal of unwanted substances from the lesion.

Effects of Oxidized Low-Density Lipoprotein on Macrophage Cytokine Production

The strong influence of cytokines and growth factors on a vast array of cell functions has led to the examination of the influence of putative in vivo regulators of the gene expression for, and protein production of, these important molecules. Of particular interest with respect to vascular disease are peptides produced by or having influence on macrophages, because cytokine influence on macrophage function may be a factor in lesion development. It has been reported that pretreatment of mouse peritoneal macrophages with oxidized LDL markedly inhibits the expression of cytokine messenger RNA that is observed on stimulation of the cells with endotoxin or maleylated albumin (Hamilton et al. 1990). Gene expression for both interleukin-1α (IL-1α) and tumor necrosis factor-α were inhibited in this system by oxidized LDL pretreatment. The inhibition was not accompanied by any significant decrease in total protein synthesis, nor was the inhibition a side effect of toxicity of the oxidized LDL. The lipid extract of the oxidized LDL contained the inhibitory substance(s). According to a recent abstract, the preceding observations were confirmed and extended to include oxidized LDL-inhibition of the gene expression for IL-1β and IL-6 (Fong et al. 1990). The extent to which these results can be generalized to other cytokines or extrapolated to the in vivo case is unknown.

Oxidized Low-Density Lipoproteins and Endothelial Cells

Endothelial Cell Oxidation of Low-Density Lipoproteins

Endothelial cells in culture, both from human umbilical vein and from a rabbit arterial endothelial cell line, were the first cells shown to oxidize LDL (Morel et al. 1984; Steinbrecher 1984). The two reports of this finding explained the mechanism by which endothelial cells modified LDL (which had been reported earlier in Henriksen et al. 1981), and thus linked the toxicity that had been reported for oxidized LDL (Hessler et al. 1983; Morel et al. 1983) and the acetylated LDL receptor recognition (Henriksen et al. 1981) to a common oxidative event. The mediators of the oxidation of LDL by endothelial cells are somewhat uncertain because of seemingly conflicting reports. The ambiguities include the finding that not all sources of endothelial cells are equally potent in their capacity to oxidize the lipoproteins. For example, the preceding two endothelial cell sources are more potent than bovine aortic endothelial cells in similar medium (Morel et al. 1984). That lipoxygenase activity is required for the oxidation to proceed is suggested by the report that ETYA, but not cyclooxygenase inhibitors, and piriprost, a nonantioxidant lipoxygenase inhibitor, both block LDL oxidation by endothelial cells (Parthasarathy et al. 1989). In a contradictory report, however, ETYA at similar concentrations failed to inhibit endothelial cell modification of the lipoprotein (van Hinsberg et al. 1986). Furthermore, there are detailed studies reporting both that SOD does (Steinbrecher 1988) and does not (Parthasarthy et al. 1989) interfere with LDL oxidation by endothelial cells. Although it is clear that under particular conditions endothelial cells can markedly enhance LDL oxidation, the cellular participants in the reaction remain uncertain.

Effects of Oxidized Low-Density Lipoprotein on Endothelial Cells

Current data support the existence of oxidized lipoproteins both in plasma and in arterial intima under particular pathologic situations; thus, the effect that oxidized LDL has on endothelial cells is particularly important. Because confluent endothelium has been shown to express the scavenger receptor (Sanan et al. 1985), there is the potential for receptor-enhanced interactions between oxidized LDL and endothelium. Clearly, however, not all of the effects on endothelium mediated by oxidized LDL are due to scavenger-receptor recognition of the ligand. It was demonstrated recently that LDL oxidized to a moderate extent, such that it was still recognizable by the LDL receptor, was able to induce the gene expression for, and production of, granulocyte-macrophage colony-stimulating factor (GM-CSF), macrophage CSF (M-CSF), and granulocyte CSF (G-CSF) in endothelial cells, even at oxidized LDL concentrations as low as 1 to 5 μg LDL protein/ml (Rajavashisth et al. 1990). The effects were observed in aortic endothelial cells from humans and rabbits, but not in human umbilical vein endothelial cells (Rajavashisth et al. 1990). Higher concentrations of LDL oxidized to a similar degree were shown earlier to inhibit markedly the production by endothelial cells of platelet-derived growth factor (PDGF) (Fox et al. 1987). The latter effect was mediated by low concentrations of acetylated LDL at even lower levels of oxidation, suggesting that the receptor recognition of acetylated LDL by the confluent endothelial cells increased the efficiency of the inhibitor(s). The inhibitory activity in oxidized LDL was extractable in the lipid phase of the lipoprotein, indicating that the apolipoprotein was not required for the inhibition. Interestingly, PDGF production by human monocyte-derived macrophages in vitro was also shown recently to be inhibited by oxidized LDL (Malden et al. 1990). These changes in growth factor production could potentially influence many of the cellular responses in the artery wall. CSFs influence the migration and proliferation of macrophages and granulocytes; G-CSF and GM-CSF alter the migration and proliferation of endothelial cells; and PDGF stimulates the migration and proliferation of smooth muscle cells.

Other potent effects of oxidized LDL on endothelium include the induction of monocyte binding to endothelial cells (Berliner et al. 1990) and the recent finding that oxidized LDL facilitates the release of endothelin from endothelial cells (Boulanger et al. 1990). The toxicity of LDL to endothelial cells was reported over a decade ago (Hessler et al. 1979; Henriksen et al. 1979b), but it was later that the modification of LDL leading to its transformation into a toxin was identified as a result of free-radical oxidation (Hessler et al. 1983; Morel et al. 1983). The toxic effects of oxidized LDL on endothelial cells can be generalized to a wide variety of cells; these effects are discussed below.

Cellular Injury by Oxidized Low-Density Lipoprotein

The toxicity of oxidized LDL has been reported to affect endothelial cells, vascular smooth muscle cells, dermal fibroblasts, and stimulated lymphocytes, among others (Hessler et al. 1979, 1983; Morel et al. 1983; Henriksen et al. 1979b; Schuh et al. 1978). Interesting traits that have been reported include the inhibition of the toxicity by HDL (Hessler et al. 1979, 1983; Henriksen et al. 1979a), the increased vulnerability of fibroblasts (and possibly other cells) to oxidized LDL cytotoxicity during the DNA synthesis phase of the cell cycle (Kosugi et al. 1987), and the ability of glutathione to inhibit the toxic effect on endothelial cells (Kuzuya et al. 1989).

That lipoprotein oxidized in vivo is cytotoxic is suggested by the report that an oxidized VLDL + LDL fraction of diabetic rat plasma is both oxidized and toxic to cultured cells and that treatment of these streptozotocin diabetic rats with the lipophilic antioxidants, vitamin E or probucol, inhibits the lipoprotein oxidation and its toxicity (Morel and Chisolm 1989). Neither antioxidant affected the hyperglycemia in these animals. Left unanswered is the question of whether blocking the formation of these cytotoxic modified lipoproteins inhibits the tissue damage that accompanies diabetes in this experimental model.

The numerous modes of oxidizing LDL, including long-term storage to allow oxidation, augmentation of the process by the addition of iron on copper ions, facilitation of the process by free radicals produced from vascular smooth muscle cells, endothelial cells or activated monocytes, all appear to render LDL cytotoxic to growing cells in culture. The cytotoxic effect does not require the LDL receptor, because fibroblasts from a homozygous familial hypercholesterolemic human, devoid of LDL receptor expression, are susceptible to oxidized LDL-induced toxicity (Henriksen et al. 1979b; Børsum et al. 1982) and because the lipid extract of oxidized LDL contains most of the cytotoxic activity (Hessler et al. 1983). LDL oxidized to an extent referred to by some as "minimally modified," as measured by thiobarbituric acid reactivity, is also cytotoxic (Hessler et al. 1983; Morel et al. 1983). It is possible that oxidized LDL cytotoxicity could be enhanced by LDL receptor uptake in instances in which the LDL is not oxidized sufficiently to block receptor recognition.

The oxidized lipid(s) that are responsible for the cytotoxicity of oxidized LDL has (have) not been identified, but there are numerous candidates, known to be formed during LDL oxidation, which have been shown to be toxic in other contexts. Oxidized derivatives of cholesterol, many of which are potent toxins, are formed in oxidized LDL. Aldehydic derivatives of fatty acid oxidation, such as the known toxin, 4-hydroxynonenal, are likewise formed (Esterbauer et al. 1987). Lysophosphatidylcholine, which is formed during LDL oxidation because of phospholipase A_2 activity intrinsic to the lipoprotein, has also been reported to damage cells.

The extent to which these injurious agents play a role in pathologic conditions in vivo is unknown. The antioxidant status of the cells, the accompanying presence of HDL, and the proliferative status of cells, could all influence their susceptibility. The toxic phenomenon has attracted interest because of the long-standing suspicion that endothelial cell injury is an event associated with atherosclerotic lesion development (Ross, 1986) and the long-standing observation that dead cell debris is a prominent component of advanced vascular lesions.

Oxidized Low-Density Lipoprotein in Vivo

Evidence for the Existence of Oxidized Low-Density Lipoprotein in Vivo

The data indicating that oxidized forms of LDL exist in vivo are compelling. The evidence suggests the presence of oxidized LDL both in arterial lesions and in plasma. The circumstances under which these data are collected warrant a caution regarding the definition of oxidized LDL. Distinctions have yet to be made in vivo between LDL that has been subjected directly to oxidative free radicals and LDL that has acquired lipid peroxidation products by exchange from adjacent cell membranes, interstitial sites, or other lipoproteins. The evidence related immediately below applies to LDL in vivo that is similar to LDL oxidized in vitro, but this evidence does not preclude that altered LDL having these characteristics is formed by a mechanism unlike the direct free radical reactions typically used to oxidize the lipoprotein in vitro.

For many years, the LDL-like particles extracted from atherosclerotic lesions of humans and animals have been recognized as different from plasma LDL (Morton et al. 1986; Smith and Slater 1972). Lipid composition differences and differences in electrophoretic mobility were among the distinguishing characteristics. More recently, however, further similarities have been identified between "LDL" extracted from lesions and oxidized LDL. For the LDL-like lipoprotein extracted from the WHHL rabbit, for example, thiobarbituric acid reactivity is enhanced compared with that of the plasma LDL (Daugherty et al. 1988), and recognition by macrophages is also increased (Ylä-Herttuala et al. 1989). The latter has also been demonstrated for the LDL extracted from human atherosclerotic lesions (Ylä-Herttuala et al. 1989; Palinski et al. 1989).

In addition to these similarities, antibodies that do not recognize LDL, but rather recognize oxidized LDL as well as LDL chemically modified by reaction with certain lipid peroxidation products, also recognize epitopes in the atherosclerotic lesions of WHHL rabbits and humans. The antibodies used in these studies include both monoclonal and polyclonal antibodies to malondialdehyde-linked LDL, 4-hydroxynonenal-modified LDL and LDL oxidized by cupric ion (Palinski et al. 1989, 1990; Haberland et al. 1988; Rosenfeld et al. 1990; Boyd et al. 1989; Mowri et al. 1988). Antibodies to malondialdehyde-LDL and hydroxynonenal-LDL cross react with malondialdehyde and 4-hydroxynonenal (respectively) linked to the lysyl residues of proteins other than LDL. The localization of these epitopes in atherosclerotic lesions has shown that the modified proteins co-localize with apolipoprotein B (Haberland et al. 1988) or in patterns adjacent to the foam cells of the lesions (Rosenfeld et al. 1990). These results are consistent with the local oxidation by macrophages of LDL that accumulates in lesions. Because antibodies recognizing oxidized LDL at unknown epitopes may, like the antibodies to malondialdehyde-lysine and hydroxynonenal-lysine mentioned previously, bind to yet other lipid peroxidation products linked to particular amino acid residues of proteins other than LDL, the results with these antibodies, viewed in the absence of other corroborative data, must be regarded as strongly suggestive of, but not proving the existence of, oxidized LDL in lesions.

Furthering the concept that oxidized LDL exists in vivo is the striking finding of circulating antibodies in the plasma of normal (asymptomatic) humans that recognize oxidized LDL and, more generally, proteins linked to malondialdehyde (Palinksi et al. 1989). This finding suggests that the oxidation of LDL or the modification of proteins by lipid peroxidation products may occur in "normalcy" and evoke an autoimmune-like response.

There is also evidence that oxidized LDL exists in plasma. In normal men, a subpopulation of LDL particles, accounting for a few percent of the total LDL mass, has been partially characterized and shown to have properties similar to those of oxidized LDL (Avogaro et al. 1988). These properties include reduced contents of vitamin E, increased density, and elevated thiobarbituric acid reactivity. Further studies are warranted to determine if this subfraction has indeed been formed by oxidation and whether it is of a normal or pathologic nature.

Oxidized lipoproteins have been demonstrated in the plasma of rats made hyperglycemic by injection of streptozotocin (Morel and Chisolm 1989). The VLDL plus LDL fraction from these animals contained marked elevations in thiobarbituric acid reactivity and, like LDL or VLDL oxidized in vitro, was toxic to proliferating cultured cells. The lipophilic antioxidants, vitamin E and probucol, administered either by interoperitoneal injection or orally, were effective in reducing both the oxidation and the toxicity, but did not affect the hyperglycemia. Insulin, on the other hand, normalized the hyperglycemia and prevented the lipoprotein oxidation and cytotoxicity. This experimental model may offer the opportunity to determine if oxidized plasma lipoproteins can damage tissue in vivo. In addition, it serves as evidence that oxidized lipoproteins can exist in plasma, a concept that was previously believed unlikely because of earlier studies showing that endothelial cell-modified LDL was cleared quickly from plasma (Nagelkerke et al. 1984). A more accurate view appears to be that the rapidity of removal of oxidized LDL from plasma is dependent upon the degree of its oxidative modification (Steinbrecher et al. 1987).

Circumstantial Evidence for Oxidized Low-Density Lipoprotein Participation in Disease

Atherosclerosis

There is no direct evidence that oxidized LDL causes atherosclerosis or other maladies. There is, on the other hand, a growing body of circumstantial evidence suggesting that this is the case. This evidence includes the fact that oxidized lipoproteins can exist in vivo in association with certain disease states. The in vitro findings indicating that oxidized LDL has properties conducive to lesion formation—enhanced macrophage recognition, cytotoxicity, and participation in monocyte recruitment—are also consistent with an atherogenic role.

Perhaps most intriguing are the demonstrations by several laboratories that probucol reduces foam-cell formation in the arteries of the WHHL rabbit (Carew et al. 1987; Kita et al. 1987) and the indication of reduced macrophage incorporation of LDL in the lesions of these treated animals (Carew et al. 1987). The cholesterol-lowering effect of probucol in one of these studies was matched by a group of animals treated with lovastatin, a cholesterol-lowering drug that reduces cholesterol synthesis (Carew et al. 1987). Thus, the simplest explanation for the antiatherogenic effect is that probucol decreased the susceptibility of LDL to oxidation and thereby inhibited its modification to become a ligand for macrophage scavenger receptors. Supporting this explanation are studies showing that LDL from humans treated with probucol is less easily oxidized in vitro than LDL from untreated humans (Parthasarathy et al. 1986b) and studies showing that probucol does not otherwise interfere with macrophage uptake of ligands for the scavenger receptor (Nagano et al. 1989).

These data suggest that antioxidants may prevent LDL from becoming atherogenic; antioxidants have been shown both to succeed and to fail at inhibiting atherogenesis in experimental models (Stein et al. 1989; Donaldson 1982; Morrissey and Donaldson 1979; Daughtery et al. 1989; Chisolm 1991). Some of the failure might be attributable to inadequate doses or to the fact that in many experimental models the atherogenic lipoprotein is the cholesterol-diet induced β-VLDL, which can be taken up by macrophages without a requirement for modification by oxidation. It is important to note that inverse correlations between plasma vitamin E levels and risk of heart disease in diverse human populations have been reported (Gey et al. 1990). In addition, a 49% decrease was recently observed in all major vascular events in patients prescribed 50 mg of β-carotene on alternate days (Gaziano et al. 1990).

The conclusion that atherosclerosis could be prevented by antioxidant therapy would certainly be a premature one. The consequences of, and caveats to, such a direction are similar to those referred to in Chapter 8 of this volume and are worth emphasis because of the lack of clear evidence for a defined causal role for lipoprotein oxidation in atherosclerosis.

Other Diseases

The existence of oxidized lipoproteins in vivo expands the potential for the participation of free radicals in numerous pathologic conditions. Free radical damage to cells is suspected of playing a role in numerous disease processes. Often the relatively short half-lives of free radicals influence the way free radical-induced cytotoxicity is regarded; for example, the target cell is often viewed as needing to be adjacent to the free radical-generating cell that injures it. The concept that free radicals can alter lipoproteins and produce more stable toxins that can travel from the free radical source to a distant target site enhances the potential influence of free radicals (Cathcart et al. 1985). Thus, any disease in which inflammatory, free radical-producing cells are activated may involve the participation of oxidized lipoproteins that can carry to various tissues oxidized lipids that markedly alter cell function. Free radical participation in carcinogenesis (see Chapter 8) may involve participation of the altered lipids of oxidized lipoproteins.

Because the free radical-alteration of lipoproteins generates through lipid peroxidation water-soluble products such as malondialdehyde that can readily alter proteins, one can speculate that this alteration of "self" protein mediated by the actions of free radicals on lipoproteins could induce an autoimmune response. Diseases like diabetes, which are characterized by both an accompanying oxidative component and autoimmune characteristics, could involve such a mechanism.

SUMMARY AND FUTURE DIRECTIONS

Existing data support the existence of oxidized lipoproteins in human and animal atherosclerotic lesions and in the plasma of diabetic rats; however, any causal role of these lipoproteins in disease awaits proof. The outcomes of experiments using the lipophilic antioxidant probucol to impede fatty streak formation in the WHHL rabbit are certainly consistent with theories proposing a role for oxidized lipoproteins in lesion formation (Steinberg et al. 1989). More data are needed, however, to show that the reduction in the early rabbit lesions is not due to another, as yet undefined, action of the drug. Studies are underway to examine the effect of probucol on the development of human arterial lesions (Regnström et al. 1990).

The diverse and numerous cellular changes effected by oxidized LDL are perhaps not surprising when viewed with the recognition that so many new compounds are produced by the oxidation. An exciting outgrowth of this heterogeneity is the anticipation that specific lipid derivatives of lipoprotein oxidation are linked to specific alterations in cell function, inviting speculation and further research into an expanded role of oxidized lipids in physiologic cell regulation in addition to their roles in pathologic conditions such as atherosclerosis.

If a theory for a causal role for oxidized LDL in atherosclerosis is proved in some form, the opportunities for intervention in the disease will certainly expand. Other than lowering LDL cholesterol and thereby decreasing the substrate for the oxidative production of atherogenic substances, lipophilic antioxidants that reside in the lipoprotein may be able to impede its modification. In addition, once the cells are identified that mediate the oxidation in vivo and once the mechanism for the oxidation is understood, inhibitors of the process may be sought. One could speculate, for example, that lipoxygenase inhibition or inhibition of local superoxide anion production might inhibit oxidation of interstitial LDL by the phagocytic macrophages of the early foam-cell lesion. Substances that selectively interfere with the uptake by macrophages of altered lipoproteins that are recognized by the scavenger receptor, or substances that block the effects of the cytotoxic oxidized lipids residing in oxidized LDL would provide alternate possibilities for exploration.

REFERENCES

Arai, H., Kita, T., Yokode, M., Narumiya, S., and Kawai, C. (1989). Multiple receptors for modified low density lipoproteins in mouse peritoneal macrophages: different uptake mechanisms for acetylated and oxidized low density lipoproteins. *Biochem. Biophys. Res. Commun.*, 159:1375–82.

Avogaro, P., Bon, G. B., and Cazzolato, G. (1988). Presence of a modified low density lipoprotein in humans. *Arteriosclerosis*, 8:79–87.

Berliner, J. A., Territo, M. C., Sevanian, A., Ramin, S., Kim, J. A., Bamshad, B., et al. (1990). Minimally modified LDL stimulates monocyte endothelial interactions. *J. Clin. Invest.*, 85:1260–66.

Biesbroeck, R., Oram, J. F., Albers, J. J., and Bierman, E. L. (1983). Specific high-affinity binding of high density lipoproteins to cultured human skin fibroblasts and arterial smooth muscle cells. *J. Clin. Invest.*, 71:525–39.

Børsum, T., Henriksen, T., Carlander, B., and Reisvaag, A. (1982). Injury to human cells in culture induced by low-density lipoprotein—An effect independent of receptor binding and endocytotic uptake of low density lipoprotein. *Scand. J. Clin. Lab. Invest.*, 42:75–81.

Boulanger, C., Hahn, A. W. A., and Lüscher, T. F. (1990). Oxidized low-density lipoproteins release endothelin from the human and porcine endothelium. *Circulation*, 82:III–224 (abstract).

Boyd, H. C., Gown, A. M., Wolfbauer, G., and Chait, A. (1989). Direct evidence for a protein recognized by a monoclonal antibody against oxidatively modified LDL in atherosclerotic lesions from a Watanabe heritable hyperlipidemic rabbit. *Am. J. Pathol.*, 135:815–25.

Bratzler, R. L., Chisolm, G. M., Colton, C. K., Smith, K. A., and Lees, R. S. (1977). The distribution of labeled low-density lipoproteins across the rabbit thoracic aorta *in vivo*. *Atherosclerosis*, 28:289–307.

Brown, M. S., and Goldstein, J. L. (1983). Lipoprotein metabolism in the macrophage implications for cholesterol deposition in atherosclerosis. *Ann. Rev. Biochem.*, 52:223–61.

Brown, M. S., and Goldstein, J. L. (1990). Scavenging for receptors. *Nature*, 343:508–09.

Caplan, B. A., and Schwartz, C. J. (1973). Increased endothelial cell turnover in areas of *in vivo* Evans Blue uptake in the pig aorta. *Atherosclerosis*, 17:401–17.

Carew, T. E., Schwenke, D. C., and Steinberg, D. (1987). Antiatherogenic effect of probucol unrelated to its hypocholesterolemic effect: evidence that antioxidants *in vivo* can selectively inhibit low density lipoprotein degradation in macrophage-rich fatty streaks and slow the progression of atherosclerosis in the Watanabe heritable hyperlipidemic rabbit. *Proc. Natl. Acad. Sci. U.SA*, 84:7725–29.

Cathcart, M. K., Chisolm, G. M., McNally, A. K., and Morel, D. W. (1988). Oxidative modification of low density lipoprotein by activated human monocytes and the cell lines U937 and HL60. *In Vitro Cell. Dev. Biol.* 24:1001–08.

Cathcart, M. K., McNally, A. K., Morel, D. W., and Chisolm, III, G. M. (1989). Superoxide anion participation in human monocyte-mediated oxidation of low density lipoprotein and conversion of low-density lipoprotein to a cytotoxin. *J. Immunol.*, 142:1963–69.

Cathcart, M. K., McNally, A. K., and Chisolm, G. M. (1991). Lipoxygenase-mediated transformation of human low density lipoprotein to an oxidized and cytotoxic complex. *J. Lipid Res.*, 32:63–70.

Cathcart, M. K., Morel, D. W., and Chisolm III, G. M. (1985). Monocytes and neutrophils oxidize low density lipoproteins making it cytotoxic. *J. Leukoc. Biol.*, 38:341–50.

Chisolm, G. M. (1991). Antioxidants and atherosclerosis: a current assessment. *Clin. Cardiol.*, 14:I-25–30.

Cushing, S. D., Berliner, J. A., Valente, A. J., Territo, M. C., Navab, M., Parhami, F., et al. (1990). Minimally modified low density lipoprotein induces monocyte chemotactic protein (MCP-1) in human endothelial smooth muscle cells. *Proc. Natl. Acad. Sci. USA*, 87:5134–38.

Daugherty, A., Zwiefel, B. S., Sobel, B. E., and Schonfeld, G. (1988). Isolation of low density lipoprotein from atherosclerotic vascular tissue of Watanabe heritable hyperlipidemic rabbits. *Arteriosclerosis*, 8:768–77.

Daugherty, A., Zweifel, B. S., and Schonfeld, G. (1989). Probucol attenuates the development of aortic atherosclerosis in cholesterol-fed rabbits. *Br. J. Pharmacol.*, 98:612–18.

DeWhalley, C. A., Rankin, S. M., Hoult, J. R. S., Jessup, W., and Leake, D. S. (1990). Flavoids inhibit the oxidative modification of low density lipoproteins by macrophages. *Biochem. Pharmacol.*, 39:1743–50.

Donaldson, W. E. (1982). Atherosclerosis in cholesterol-fed Japanese quail: evidence for amelioration by dietary vitamin E. *Poultry Science*, 61:2097–2102.

Esterbauer, H., Jurgens, G., Quehenberger, O., and Koller, E. (1987). Autoxidation of human low density lipoprotein: loss of polyunsaturated fatty acids and vitamin E and generation of aldehydes. *J. Lipid Res.*, 28:495–509.

Fong, L. G., Parthasarathy, S., Witztum, J. L., and Steinberg, D. (1987). Nonenzymatic oxidative cleavage of peptide bonds in apoprotein B_{100}. *J. Lipid Res.*, 28:1466–77.

Fong, L. G., Fong, T. A. T., and Cooper, A. D (1990). Inhibition of macrophage interleukin-1-β mRNA expression by oxidized-LDL. *Circulation*, 82:III-207 (abstract).

Fox, P. L., Chisolm, G. M., and DiCorleto, P. E. (1987) Lipoprotein-mediated inhibition of endothelial cell production of platelet-derived growth factor-like protein depends on free radical lipid peroxidation. *J. Biol. Chem.*, 262:6046–54.

Freeman, M., Ekkel, Y., Rohrer, L., Penman, M., Freedman, N., Chisolm, G. M.,et al. (1991). Expression of type I and type II bovine scavenger receptors in Chinese hamster ovary cells: lipid droplet accumulation and nonreciprocal cross competition by acetylated and oxidized LDL. *Proc. Natl. Acad. Sci. USA.*, 88:4931–35.

Gaziano, J. M., Manson, J. E., Ridker, P. M., Buring, J. E., and Hennekens, C. H. (1990). Beta carotene therapy for chronic stable angina. *Circulation*, 82:III–201 (abstract).

Gerrity, R. G. (1981). The role of the monocyte in atherogenesis. I. Transition of blood-borne monocytes into foam cells in fatty lesions. *Am. J. Pathol.*, 103:181–90.

Gey, K. F., Puska, P., Jordan, P., and Moser U. K. (1990). Inverse correlation between plasma vitamin E and mortality from ischemic heart disease in cross-cultural epidemiology. *Am. J. Clin. Nutr.*, 53:326S–34S.

Goldstein, J. L., and Brown, M. S. (1977). The low-density lipoprotein pathway and its relation to atherosclerosis. *Ann. Rev. Biochem.*, 46:897–930.

Goldstein, J. L., Hoff, H. F., Ho, Y. K., Basu, S. K., and Brown, M. S. (1981). Stimulation of cholesteryl ester synthesis in macrophages by extracts of atherosclerotic human aortas and complexes of albumin/cholesteryl esters. *Arteriosclerosis*, 1:210–26.

Haberland, M. E., Fogelman, A. M., and Edwards, P. A. (1982). Specificity of receptor mediated recognition of malondialdehyde modified low density lipoproteins. *Proc. Natl. Acad. Sci. USA*, 79:1712–16.

Haberland, M., Fong, D., and Cheng, L. (1988). Malondialdehyde-altered protein occurs in atheroma of Watanabe heritable hyperlipidemic rabbits. *Science*, 241:215–18.

Hamilton, T. A., Ma, G., and Chisolm, G. M. (1990). Oxidized low density lipoprotein suppresses the expression of tumor necrosis factor-alpha mRNA in stimulated murine peritoneal macrophages. *J. Immunol.*, 144:2343–50.

Heinecke, J. W., Rosen, H., and Chait, A. (1984). Iron and copper promote modification of low density lipoprotein by human arterial smooth muscle cells in culture. *J. Clin. Invest.*, 74:1890–94.

Henriksen, T., Evensen, S. A., and Carlander, B. (1979a). Injury to cultured endothelial cells induced by low density lipoproteins: protection by high density lipoproteins. *Scand. J. Clin. Lab. Invest.*, 39:369–75.

Henriksen, T., Evensen, S. A., and Carlander, B. (1979b). Injury to human endothelial cells in culture induced by low density lipoproteins. *Scand. J. Clin. Lab. Invest.*, 39:361–68.

Henriksen, T., Mahoney, E. M., and Steinberg, D. (1981). Enhanced macrophage degradation of low density lipoprotein previously incubated with cultured endothelial cells: recognition by the receptor for acetylated low density lipoproteins. *Proc. Natl. Acad. Sci. USA*, 78:6499–6503.

Hessler, J. R., Morel, D. W., Lewis, L. J., and Chisolm, G. M. (1983). Lipoprotein oxidation and lipoprotein-induced cytotoxicty. *Arteriosclerosis*, 3:215–22.

Hessler, J. R., Robertson, Jr., A. L., and Chisolm, G. M. (1979). LDL-induced cytotoxicity and its inhibition by HDL in human vascular smooth muscle and endothelial cells in culture. *Atherosclerosis*, 32:213–29.

Hiramatsu, K., Rosen, H., Heinecke, J. W., Wolfbauer, G., and Chait, A. (1987). Superoxide initiates oxidation of low density lipoprotein by human monocytes. *Arteriosclerosis*, 7:55–60.

Hoff, H. F., and Gaubatz, J. W. (1982). Isolation, purification and characterization of a lipoprotein containing apo B from the human aorta. *Atherosclerosis*, 42:273–97.

Hoff, H. F., Heideman, C. I., Gaubatz, V. W., Scott, D. W., and Gotto, A. M. (1978). Detergent extraction of tightly-bound apo-B from extracts of normal aortic intima and plaques. *Exp. Mol. Pathol.*, 28:290–300.

Jessep, W., Jurgens, G., Lang, J., Esterbauer, H., and Dean, R. T. (1986). Interaction of 4-hydroxynonenal-modified low-density lipoproteins with the fibroblast apolipoprotein B/E receptor. *Biochem. J.*, 234:245–48.

Jurgens, G., Hoff, H. F., Chisolm, G. M., and Esterbauer, H. (1987). Modification of human serum low density lipoprotein by oxidation—characterization and pathophysiologic implications. *Chem. Phys. Lipids*, 45:315–36.

Kita, T., Nagano, Y., Yokode, M., Ishii, K., Kume, N., Ooshima, A., et al. (1987). Probucol prevents the progression of atherosclerosis in Watanabe heritable hyperlipidemic rabbit, an animal model for familial hypercholesterolemia. *Proc. Natl. Acad. Sci. USA*, 84:5928–31.

Kodama, T., Freeman, M., Rohrer, L., Zabrecky, J., Matsudaira, P., and Kreiger, M. (1990). Type I macrophage scavenger receptor contains α-helical and collagen-like coiled coils. *Nature*, 343:532–35.

Kosugi, K., Morel, D. W., DiCorleto, P. E., and Chisolm, G. M. (1987). Toxicity of oxidized low density lipoprotein to cultured fibroblasts is selective for the S phase of the cell cycle. *J. Cell. Physiol.*, 102:119–27.

Kowal, R. C., Herz, J., Goldstein, J. L., Esser, V., and Brown, M. S. (1989). Low density lipoprotein receptor-related protein mediates uptake of cholesteryl esters derived from apoprotein E-enriched lipoproteins. *Proc. Natl. Acad. Sci. USA*, 86:5810–14.

Kuzuya, M., Naito, M., Funaki, C., Hayashi, T., Asai, K., and Kuzuya, F. (1989). Protective role of intracellular glutathione against oxidized low density lipoprotein in cultured endothelial cells. *Biochem. Biophys. Res. Commun.* 163:1466–72.

McNally, A. K., Chisolm, III, G. M., Morel, D. W., and Cathcart, M. K. (1990). Activated human monocytes oxidize low-density lipoprotein by a lipoxygenase-dependent pathway. *J. Immunol.*, 145:254–59.

Malden, L. T., Ross, R., and Chait, A. (1990). Oxidatively modified low density lipoproteins inhibit expression of platelet derived growth factor by human monocyte-derived macrophages. *Clin. Res.*, 38:291A (abstract).

Morel, D. W., and Chisolm, G. M. (1989). Antioxidant treatment of diabetic rats inhibits lipoprotein oxidation and cytotoxicity. *J. Lipid Res.*, 30:1827–34.

Morel, D. W., DiCorleto, P. E., and Chisolm, G. M. (1984). Endothelial and smooth muscle cells alter low density lipoprotein *in vitro* by free radical oxidation. *Arteriosclerosis*, 4:357–64.

Morel, D. W., Hessler, J. R., and Chisolm, G. M. (1983). Low density lipoprotein cytotoxicity induced by free radical peroxidation of lipid. *J. Lipid Res.*, 24:1070–76.

Morrissey, R. B., and Donaldson, W. E. (1979). Cholesteremia in Japanese quail: response to a mixture of vitamins C and E and choline chloride. *Artery*, 5:182–92.

Morton, R. E., West, G. E., and Hoff, H. F. (1986). A low density lipoprotein-sized particle isolated from human atherosclerotic lesions is internalized by macrophages via a non-scavenger-receptor mechanism. *J. Lipid Res.*, 27:1124–34.

Mowri, H., Ohkuma, S., and Takano, T. (1988). Monoclonal DLR1a/104G antibody recognizing peroxidized lipoproteins in atherosclerotic lesions. *Biophys. Biochim. Acta*, 963:208–14.

Nagano, Y., Kita, T., Yokode, M., Ishii, K., Kume, N., Otani, H., Arai, H., and Kawai, C. (1989). Probucol does not affect lipoprotein metabolism in macrophages of Wantanabe heritable hyperlipidemic rabbits. *Arteriosclerosis*, 9:453–61.

Nagelkerke, J. F., Havekes, J., van Hinsbergh, V. W., and van Berkel, T. J. (1984). In vivo catabolism of biologically modified LDL. *Arteriosclerosis*, 4:256–94.

Palinski, W., Rosenfeld, M. E., Ylä-Herttuala, S., Gurtner, G. C., Socher, S. S., Butler, S. W., et al. (1989). Low density lipoprotein undergoes oxidative modification in vivo. *Proc. Natl. Acad. Sci. USA*, 86:1372–76.

Palinski, W., Ylä-Herttuala, S., Rosenfeld, M. E., Butler, S. W., Socher, S. A., Parthasarathy, S., et al. (1990). Antisera and monoclonal antibodies specific for epitopes generated during oxidative modification of low density lipoprotein. *Arteriosclerosis*, 10:325–35.

Parathasarathy, S. (1987). Oxidation of low-density lipoprotein by thiol compounds leads to its recognition by the acetyl LDL receptor. *Biochim. Biophys. Acta*, 917:337–40.

Parthasarathy, S., Printz, D. J., Boyd, D., Joy, L., and Steinberg, D. (1986a). Macrophage oxidation of low density lipoprotein generates a form recognized by the scavenger receptor. *Arteriosclerosis*, 6:505–10.

Parthasarathy, S., Steinbrecher, U. P., Barnett, J., Witztum, J. L., and Steinberg, D. (1985). Essential role of phospholipase A_2 activity in endothelial cell-induced modification of low density lipoprotein. *Proc. Natl. Acad. Sci. USA*, 82:3000–04.

Parthasarathy, S., Wieland, E., and Steinberg, D. (1989). A role for endothelial cell lipoxygenase in the oxidative modification of low density lipoprotein. *Proc. Natl. Acad. Sci. USA*, 86:1046–50.

Parthasarathy, S., Young, S. G., Witztum, J. L., Pittman, R. C., and Steinberg, D. (1986b). Probucol inhibits oxidative modification of low density lipoprotein. *J. Clin. Invest.*, 77:641–44.

Quehenberger, O., Koller, E., Jurgens, G., and Esterbauer, H. (1987). Investigation of lipid peroxidation in human low density lipoprotein. *Free Radic. Res. Commun.*, 3:233–42.

Quinn, M. T., Parthasarathy, S., and Steinberg, D. (1985). Endothelial cell-derived chemotactic activity for mouse peritoneal macrophages and the effects of modified forms of low density lipoprotein. *Proc. Natl. Acad. Sci. USA*, 82:5949–53.

Quinn, M. T., Parthasarathy, S., and Steinberg, D. (1988). Lysophosphatidylcholine: a chemotactic factor for human monocytes and its potential role in atherogenesis. *Proc. Natl. Acad. Sci. USA*, 85:2805–09.

Quinn, M. T., Parthasarathy, S., Fong, L. G., and Steinberg, D. (1987). Oxidatively modified low density lipoproteins: a potential role in recruitment and retention of monocyte/macrophages during atherogenesis. *Proc. Natl. Acad. Sci. USA*, 84:2995–98.

Rajavashisth, T. B., Andalibi, A., Territo, M C., Berliner, J. A., Navab, M., Fogelman, A. M., et al. (1990). Induction of endothelial cell expression of granulocyte and monocyte-macrophage chemotactic factors by modified low density lipoproteins. *Nature*, 344:254–57.

Regnström, J., Walldius, G., Carlson, L. A., and Nilsson, J. (1990). Effect of probucol treatment on the susceptibility of low density lipoprotein isolated from hypercholesterolemic patients to become oxidatively modified *in vitro*. *Atherosclerosis*, 82:43–51.

Rohrer, L., Freeman, M., Kodama, T., Penman, M., and Krieger, M. (1990). Coiled-coil fibrous domains mediate binding by the macrophage scavenger type II. *Nature*, 343:570–72.

Rosenfeld, M. E., Palinski, W., Ylä-Herttuala, S., Butler, S., and Witztum, J. L. (1990). Distribution of oxidation specific lipid-protein adducts and apolipoprotein B in atherosclerotic lesions of varying severity from WHHL rabbits. *Arteriosclerosis*, 10:336–49.

Ross, R. (1986). The pathogenesis of atherosclerosis—an update. *N. Engl. J. Med.*, 314:488–500.

Ross, R., and Glomset, J. A. (1973). Atherosclerosis and the arterial smooth muscle cell. Proliferation of smooth muscle is a key event in the genesis of the lesions of atherosclerosis. *Science*, 180:1332–39.

Sanan, D. A., Strumpfer, A. E. M., Van der Westhuyzen, D. R., and Coetzee, G. A. (1985). Native and acetylated low density lipoprotein metabolism in proliferating and quiescent bovine endothelial cells in culture. *Eur. J. Cell. Biol.*, 36:81–90.

Schuh, J., Novogrodsky, A., and Haschemeyer, R. H. (1978). Inhibition of lymphocyte mitogenesis by autoxidized low-density lipoprotein. *Biochem. Biophys. Res. Commum.*, 84:763–68.

Schwenke, D. C., and Carew, T. E. (1989). Initiation of atherosclerotic lesions in cholesterol-fed rabbits. I. Focal increases in arterial LDL concentration precede development of fatty streak lesions. *Arteriosclerosis*, 9:895–907.

Slotte, J. P., Oram, J. F., and Bierman, E. L. (1987). Binding of high density lipoproteins to cell receptors promotes translocation of cholesterol from intracellular membranes to the cell surface. *J. Biol. Chem.*, 262:12904–07.

Smith, E. B., and Ashall, C. (1983). Low-density lipoprotein concentration in interstitial fluid from human atherosclerotic lesions. Relation to theories of endothelial damage and lipoprotein binding. *Biochim. Biophys. Acta*, 43:249–57.

Smith, E. B., and Slater, R. W. (1972). Relationship between low density lipoprotein in aortic intima and serum lipid levels. *Lancet*, 1:463–69.

Sparrow, C. P., Parthasarathy, S., and Steinberg, D. (1988). Enzymatic modification of low density lipoprotein by purified lipoxygenase plus phospholipase A_2 mimics cell-mediated oxidative modification. *J. Lipid Res.*, 29:745–53.

Sparrow, C. P., Parthasarathy, S., and Steinberg D. (1989). A macrophage receptor that recognizes oxidized low density lipoprotein but not acetylated low lipoprotein. *J. Biol. Chem.*, 264:2599–2604.

Stafforini, D. M., Prescott, S. M., and McIntyre, T. M. (1987). Human platelet activating factor acetylhydrolase: association with lipoprotein particles and role in the degradation of platelet-activating factor. *J. Biol. Chem.*, 262:4223–30.

Stein, Y., Stein, O., Delplanque, B., Fesmire, J. D., Lee, D. M., and Alaupovic, P. (1989). Lack of effect of probucol on atheroma formation in cholesterol-fed rabbits kept at comparable plasma cholesterol levels. *Atherosclerosis*, 75:145–55.

Steinberg, D., Parthasarathy, S., Carew, T. E., Khoo, J. C., and Witztum, J. L. (1989). Beyond cholesterol: modifications of low-density lipoprotein that increase its atherogenicity. *N. Engl. J. Med.*, 320:915–24.

Steinbrecher, U. P. (1987). Oxidation of human low density lipoproteins results in derivatization of lysine residues of apolipoprotein B by lipid peroxide decomposition products. *J. Biol. Chem.*, 262:3603–08.

Steinbrecher, U. P. (1988). Role of superoxide in endothelial-cell modification of low density lipoproteins. *Biochim. Biophys. Acta.*, 959:20–30.

Steinbrecher, U. P., Parthasarathy, S., Leake, D. S., Witztum, J. L., and Steinberg, D. (1984). Modification of low density lipoprotein by endothelial cells involves lipid peroxidation and degradation of low density lipoprotein phospholipids. *Proc. Natl. Acad. Sci. USA*, 83:3883–07.

Steinbrecher, U. P., and Pritchard, P. H. (1989). Hydrolysis of phosphatidylcholine during LDL oxidation is mediated by platelet-activating factor acetylhydrolase. *J. Lipid Res.*, 30:305–15.

Steinbrecher, U. P., Witztum, J. L., Parthasarathy, S., and Steinberg, D. (1987). Decrease in reactive amino groups during oxidation of endothelial cell modification of LDL. Correlation with changes in receptor-mediated catabolism. *Arteriosclerosis*, 7:135–43.

Steinbrecher, U. P., Zhang, H., and Lougheed, M (1990). Role of oxidatively modified LDL in atherosclerosis. *Free Radic. Biol. Med.*, 9:155–68.

Thomas, W. A., Florentin, R. A., Nam, S. C., Kim, D. N., Jones, R. M., and Lee, K. T. (1968). Preproliferative phase of atherosclerosis in swine fed cholesterol. *Arch. Pathol.*, 86:621–43.

van Hinsberg, V. W., Scheffer, M., Havekes, L., and Kemper, J. J. (1986). Role of endothelial cells and their products in the modification of low density lipoproteins. *Biochim. Biophys. Acta.*, 878:49–64.

Ylä-Herttuala, S., Palinski, W., Rosenfeld, M. E., Parthasarathy, S., Carew, T. E., Butler, S., et al. (1989). Evidence for the presence of oxidatively modified LDL in atherosclerotic lesions of rabbit and man. *J. Clin. Invest.* 84:1086–95.

5

DNA Damage and Stress Responses Caused by Oxygen Radicals

STUART LINN

Oxygen undergoes four successive one-electron reductions on conversion to water:

$$O_2 \rightarrow \dot{O}_2^- \rightarrow H_2O_2 \rightarrow \dot{O}H \rightarrow H_2O$$

Each of the first four species in this scheme is chemically very active, and it is significant that the presence of one of these species in the cell usually results in interconversions to the other three, as discussed below. In addition, oxygen can exist in a chemically more active state as singlet oxygen, $^1O_2$.

Oxygen radicals exist in the cell as a consequence of normal respiration in an oxygen atmosphere. Their levels can be greatly enhanced by exposure to radiation and chemicals, by diseased states, or by natural functions such as those of the neutrophils.

This chapter presents an overview of the chemical, biochemical, and molecular biologic components of oxygen radical damage to the cell and summarizes briefly our knowledge of how the cell protects itself in an active oxygen environment. Such knowledge stems largely from studies of bacteria, for which our adeptness in applying molecular genetic techniques is greatest. However, the chemistry applies to mammalian (human) cells as well, and from what we know about mammalian enzymes, the enzymology also applies.

DNA DAMAGE AND REPAIR

Perhaps the main target of oxygen radicals in terms of toxicity is DNA. Certainly, as noted below, DNA repair is a major component of the cell's protective responses. DNA damage occurs by attack of a free radical at either a sugar or a base residue (Imlay and Linn 1988). Attack at a base produces several general classes of damage (Figure 5-1): simple base modification such as the formation of 5-hydroxymethyluracil; oxidized pyrimidine residues such as thymine glycol and its breakdown products (Brimer and Lindahl 1984); oxidized purines such as 4,6-diamino-5-formadidopyrimidine (Fapy) and 8-oxyguanine or, in cases where the damage is unstable, a baseless (AP) site.

Attack that produces strand breaks can also occur at a sugar residue (Imlay and Linn 1988). These are not generally simple hydrolytic breaks such as those produced by DNases, but instead usually contain 3′-phosphate residues and possibly sugar fragments, such as a glycolate residue, or even gaps at the site of cleavage. More complex products formed by radicals include crosslinks and double-strand breaks.

In both prokaryotes and eukaryotes, double-strand breaks, DNA crosslinks and possibly other DNA damage are apparently repaired by recombinational DNA repair mechanisms, whereas simple base and sugar damage is repaired by DNA excision repair (Figure 5-2). Nucleotide excision repair is initiated by incision endonucleases that recognize abnormalities in the DNA structure. Excision is then mediated either by the same incising enzyme or by exonucleases that are capable of excising damaged nucleotides or terminal sugar residues. Terminal sugar residues or fragments can also be excised by specific phosphodiesterases (Franklin and Lindahl 1988; Johnson and Demple 1988).

Base excision repair is mediated by specific DNA glycosylases, each of which recognizes a specific type of lesion. These enzymes remove the damaged base by cleaving the base-sugar bond to yield a baseless or AP site. DNA glycosylases for thymine glycol and related pyrimidines, Fapy and related purines, and hydroxymethyluracil have been reported (Wallace 1988). The number and ubiquitous distribution of these enzymes point out the large investment that the cell must make in processing oxidative DNA damage.

Once the base is removed, a specific AP endonuclease—often associated with the DNA glycosylase—breaks the phosphodiester backbone next to the AP site, either through a hydrolytic or a β-elimination mechanism (Kim and Linn 1988). The resulting terminal lesion is then removed either by an exonuclease, a specific phosphodiesterase, or a second AP endonuclease. In any case, the resulting gap is filled by DNA polymerase and the nick is sealed with DNA ligase.

KILLING AND MUTAGENESIS BY H_2O_2

Using *Escherichia coli* and H_2O_2 as a model system in which to study oxyradical damage, we noted (Imlay and Linn 1986, 1987a, b) that 1 to 2

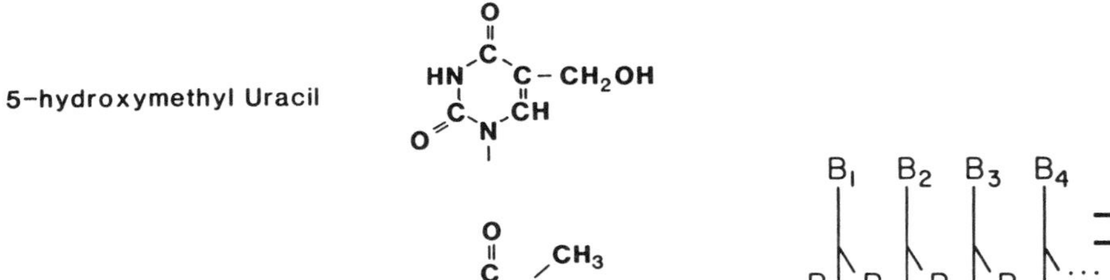

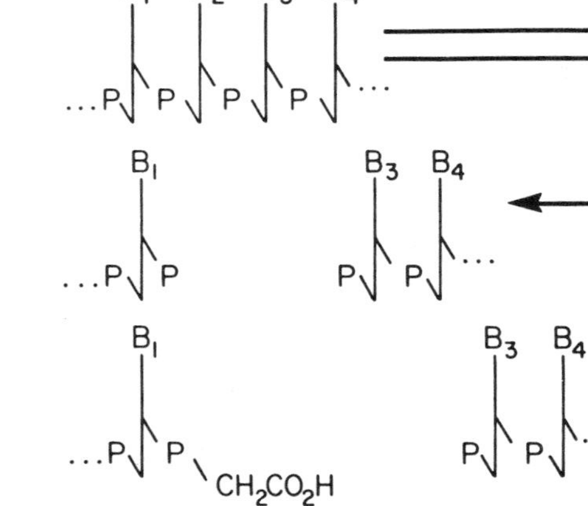

Figure 5–1 Examples of DNA damage caused by oxygen radicals. Some base simple damage such as the formation of hydroxymethyluracil is relatively unimportant. A number of pyrimidines oxidized at the 5- or 6-ring positions are formed by oxygen radicals, and these often break down to smaller base residues (Brimer and Lindahl 1984). These secondary products are generally repaired by DNA glycosylases (Wallace 1988). Likewise, purine imidazole rings are oxidized to several products, such as Fapy, which are also repaired by DNA glycosylases. 8-Oxyguanine (and 8-oxyadenine) are repaired by unknown mechanisms. Sugar damage usually results in strand breaks containing sugar residues, phosphomonoesters, or small gaps.

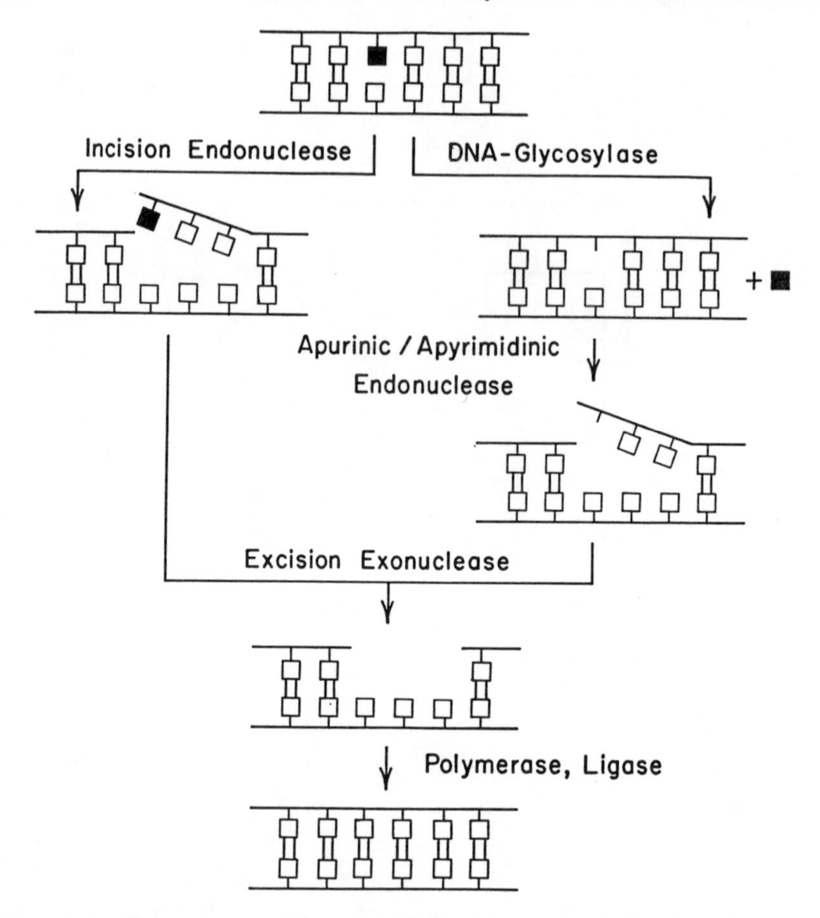

Figure 5–2 The two primary pathways of DNA excision repair. Nucleotide excision repair is mediated by incision endonucleases that cut the DNA phosphodiester backbone near the damaged nucleotide (or by a break made by the damaging agent.) Damaged material is then removed by a second endonucleolytic excision or by an exonuclease. The resulting gap is filled and sealed by DNA polymerase and ligase.

Base excision repair is mediated by removal of the damaged base by specific DNA glycosylases, then by the action of specific endonucleases that recognize the resulting baseless sugars. The terminal sugar is removed by an exonuclease, a second AP endonuclease, or a specific phosphodiesterase. The many alternative means in the same cell for excising the same damage point out the importance of DNA repair.

mM H_2O_2 is more toxic than concentrations in the range of 2 to 20 mM (Figure 5-3). We also observed that strains that are deficient in recombinational DNA repair (*recA*), excision repair (*xth*), or both types of repair (*polA*) are extremely sensitive to such killing, but the same peculiarly shaped dose-response curve is observed. This characteristic dose-response curve was observed for mutagenesis or other indicators of DNA damage; in each instance, the rate of DNA damage was two- to three-fold greater at 1 to 2 mM than at 2 to 20 mM H_2O_2. This unusual and characteristic

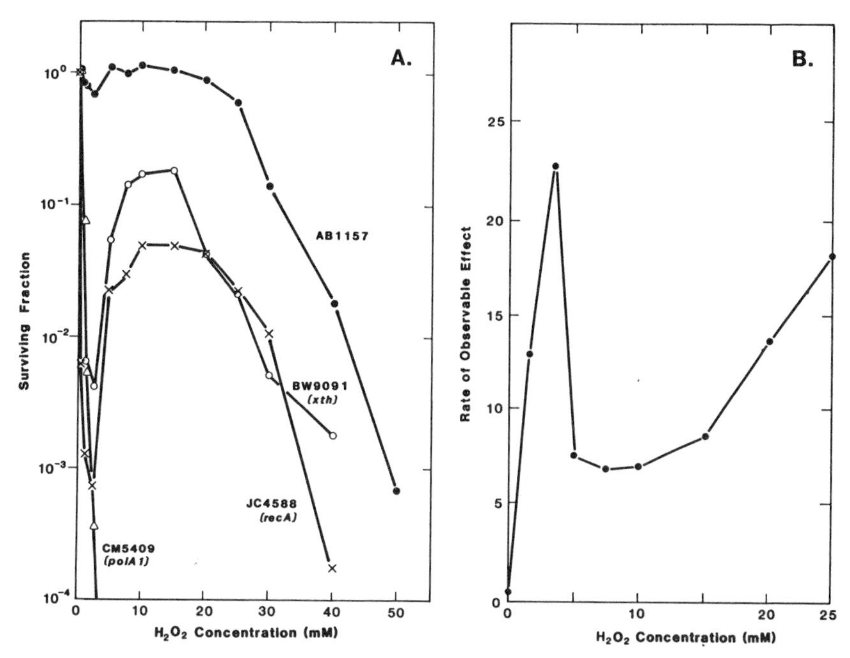

Figure 5–3 Examples of dose-response curves of H_2O_2 toxicity for *Escherichia coli*. **A.** Survival curves for a wild-type strain (AB1157), a strain deficient in DNA excision repair (BW9091), defective exonuclease III), a strain deficient in recombinational repair (JC4588, defective RecA protein), and a strain deficient in both types of repair (CM5409, defective DNA polymerase I). Exposures were for 15 minutes (Imlay and Linn 1986, 1987; Linn and Imlay 1987). **B.** A dose-response curve plotted as rate of effect (in this case, rate of mutant formation) (Linn and Imlay 1987). Similar curves are found for rate of killing, rate of phage lambda induction, and other indicators of DNA damage.

dose response provided us with a distinct hallmark with which to proceed further in the characterization of the mechanism of toxicity.

IN VIVO DNA DAMAGE REQUIRES ACTIVE METABOLISM TO GENERATE CONTINUOUS POOLS OF NADH AND AVAILABLE IRON

Among the first observations we made was that DNA damage did not occur if cells were exposed to H_2O_2 in buffer lacking a carbon source (Imlay and Linn 1986). Respiration was evidently required for the H_2O_2 toxicity to be observable. Conversely, DNA damage was enhanced in cells grown anaerobically (Imlay and Linn 1987a, b). The response of *E. coli* to anaerobic growth is *fnr* (*f*umarate *n*itrate *r*eductase), a regulon that positively controls the synthesis of fumarate, nitrate and nitrite reductases, several cytochromes and so forth, but negatively regulates *ndh*, the gene for NADH dehydrogenase II (Spiro et al. 1989). A genetic analysis of mutants of these genes allowed us to implicate the negative regulation of NADH dehydro-

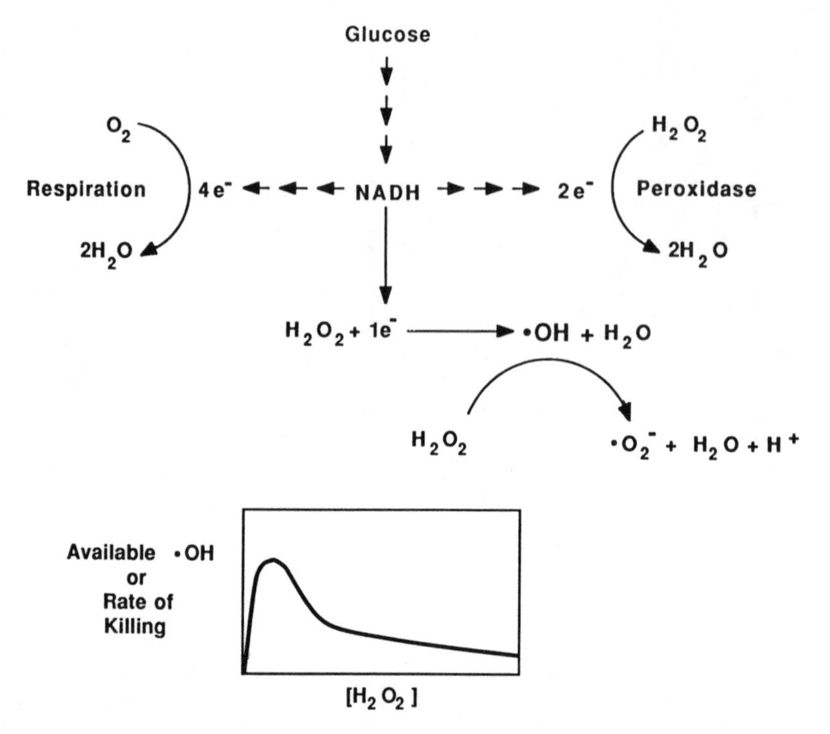

Figure 5–4 Hypothetical pathway for activation of H_2O_2 by NADH. The scheme is described in the text.

genase as the sensitizing step (Imlay and Linn 1987a; Imlay et al. 1988). NADH dehydrogenase, or diaphorase, catalyzes the reaction

$$NADH + H^+ + R \rightarrow NAD^+ + RH_2$$

NADH levels also are raised by blocking respiration with KCN, and the same sensitization to H_2O_2 ensues. Based on these results, the working model shown in Figure 5-4 was proposed (Imlay and Linn 1988; Imlay and Linn 1987b). According to the model, NADH is normally utilized in aerobic respiration ultimately to donate two electron pairs to molecular oxygen to form water. It is also a potential source of two electrons to reduce H_2O_2 to water. However, by some means, one electron can be transferred from NADH to H_2O_2 to generate $\dot{O}H$. The rate of this transfer is limited by electron flow to form NADH, however, so as to govern a maximum possible rate of $\dot{O}H$ formation. On the other hand $\dot{O}H$ can oxidize H_2O_2 to $\dot{O}_2^-$; at high H_2O_2 concentrations, the amount of $\dot{O}H$ available for cell damage is lowered.

What is the reaction agent that mediates one-electron transfers from NADH to H_2O_2? One possibility is the iron-catalyzed Fenton reaction,

$$Fe^{2+} + H_2O_2 + H^+ \rightarrow Fe^{3+} + H\dot{O} + H_2O$$

Indeed, iron chelators such as *o*-phentanthroline or dipyridyl protect against killing or mutagenesis by H_2O_2 in the 1 to 20 mM range (Imlay et al. 1988). However, various alcohol scavengers of free $\dot{O}H$ do not protect against H_2O_2 toxicity, implying that freely diffusible $\dot{O}H$ was not the damaging agent (Imlay and Linn 1988; Imlay et al. 1988).

IN VITRO FENTON REACTIONS WITH Fe^{2+} DUPLICATE THE IN VIVO DOSE RESPONSE

To study the hypothesis in greater detail, we studied the putative reactions in vitro (Imlay and Linn 1988; Imlay et al. 1988). Indeed, when available $\dot{O}H$ formed by $FeSO_4$ and H_2O_2 was monitored with an analine dye sensitive to $\dot{O}H$, the same characteristic dose-response curve was obtained: Maximal dye bleaching occurred at low concentrations of H_2O_2, and concentrations up to 30 mM reduced the available $\dot{O}H$ two- to threefold, confirming the concept that H_2O_2 competes for $\dot{O}H$. In this case, the maximal $\dot{O}H$ available was dictated by the 50 μM $FeSO_4$ present, and maximal dye bleaching occurred at approximately 200μM H_2O_2. The latter value is lower than the maximal in vivo due in part or totally to catalase, peroxidases, and so forth competing for H_2O_2 in the cell.

When killing of phage or nicking of DNA was monitored in vitro in the presence of $FeSO_4$ and H_2O_2, the same type of dose response was observed (Imlay and Linn 1988; Imlay et al. 1988). However, in the phage and DNA experiments, but not in the dye bleaching, ethanol and other alcohols did not efficiently quench the damage. Evidently, Fe^{2+} bound to DNA produces a nondiffusible $\dot{O}H$ radical that is reactive with H_2O_2 but not with the alcohols, whereas with the dye, freely diffusable $\dot{O}H$ radical was formed by Fe^{2+} in solution.

NADH DRIVES THE FENTON REACTION AND DNA NICKING IN VITRO

When DNA nicking is generated by H_2O_2 and $FeSO_4$ (Figure 5-5A), nicking occurs rapidly (probably within several seconds) and then ceases as the Fe^{2+} is oxidized. When NADH is present, however, nicking continues as the Fe^{3+} is reduced back to Fe^{2+}. When Fe^{3+} is utilized, DNA nicking occurs only when NADH is present (Figure 5-5B). In summary, NADH can drive the Fenton reaction—and the DNA-damaging reactions—by reducing Fe^{3+} to Fe^{2+}.

The scheme for NADH reduction of Fe^{3+} by a charge transfer complex was described by Gutman and colleagues (Gutman et al. 1968; Gutman and Eisenbach 1973):

$$NADH + Fe^{3+} \rightleftharpoons NADH - Fe^{3+}$$
$$NADH - Fe^{3+} \rightleftharpoons NAD\dot{H}^{+} - Fe^{2+}$$
$$Fe^{2+} - NAD\dot{H}^{+} \rightarrow Fe^{2+} + NAD\dot{H}^{+}$$
$$NAD\dot{H}^{+} \rightarrow NA\dot{D} + H^{+}$$

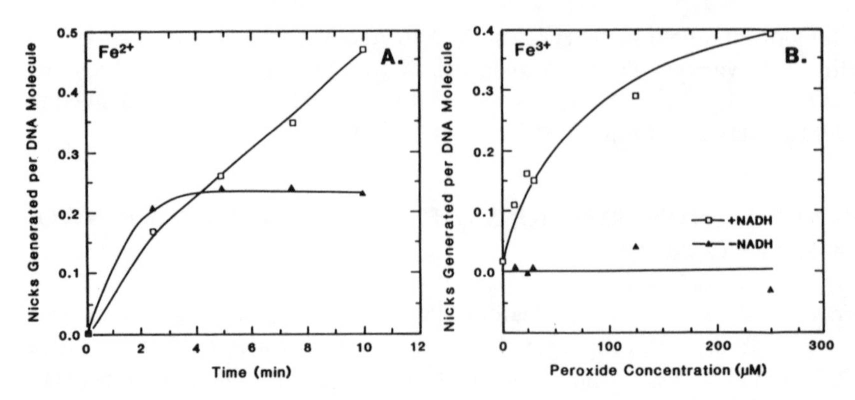

Figure 5–5 DNA damage generated in vitro by H_2O_2/FE/NADH. **A.** Reactions contained 200 μM NADH where indicated, 80 nM $FeSO_4$, 0.5 mM H_2O_2, 20 μM PM2 DNA-nucleotide, and 10 mM ethanol. **B.** Reactions contained 200 μM NADH where indicated, 2 μM $FeCl_3$, H_2O_2 as indicated, 20 μM PM2 DNA-nucleotide, and 1 mM ethanol. Incubations were for 9 minutes. (Data are taken from Chin et al., manuscript to be submitted.)

The reaction sequence is driven to completion by the essentially irreversible breakdown of $N\dot{A}DH^+$.

In conclusion, the in vitro reactions support the idea that, on exposure of cells to H_2O_2, there is a rapid Fe^{2+}-catalyzed Fenton reaction to form a nondiffusible hydroxyl radical and Fe^{3+}. This is followed in turn by the NADH-driven reduction of Fe^{3+} to allow further reaction:

$$Fe^{2+} - \text{chelate} + H_2O_2 \rightarrow Fe^{3+} - O\dot{H} - \text{chelate} + H^+ + H_2O$$

$$Fe^{3+} - O\dot{H} - \text{chelate} \rightarrow Fe^{3+} - \text{chelate} + \dot{O}H$$

$$NADH + Fe^{3+} - \text{chelate} \rightarrow \rightarrow \rightarrow Fe^{2+} - \text{chelate} + NA\dot{D} + H^+$$

The iron chelator(s) in vivo and the fate of $NA\dot{D}$ are unknown and probably vary with the cell type and state. Also unknown in detail are the intracellular interconversion reactions between H_2O_2, O_2, and $\dot{O}_2^-$.

CONSTITUTIVE PROTECTION MECHANISMS FOR OXYGEN RADICAL TOXICITY

The cell has two apparent lines of defense against DNA damage by oxygen radicals: DNA repair and scavengers of the radicals. These defenses are apparent both in constitutive and in induced, stress-response situations.

If one looks at the constitutive protection mechanisms, one sees that the cell has efficient means for DNA repair by excision or recombinational pathways. In *E. coli*, for which DNA repair is best understood, several alternative enzymes apparently exist for repairing each type of lesion formed by oxygen radicals (Imlay and Linn 1988; Wallace 1988).

When one turns to the scavenging protection strategies, one finds that superoxide dismutase (SOD) is probably the most important factor. In *E. coli*, for example, mutants lacking SOD are far more sensitive to H_2O_2 than those lacking catalase (Imlay and Linn 1987a). SOD catalyzes the reaction

$$2\dot{O}_2^- + 2H^+ \rightarrow H_2O_2 + O_2$$

Superoxide itself is relatively stable, and the cell might exploit its stability for the formation of a free radical sink in the form of superoxide, which can be tolerated as long as SOD is available. Superoxide is formed by the action of various exogenous redox cycling agents such as paraquat. In addition, it can be formed by several reactions alluded to earlier:

$$H\dot{O} + H_2O_2 \rightarrow \dot{O}_2^- + H_2O + H^+$$

or

$$Fe^{3+} - O\dot{H} - \text{chelate} + H_2O_2 \rightarrow Fe^{3+} - \text{chelate} + \dot{O}_2^- + H_2O + H^+$$

In addition, we have noted (Imlay and Linn 1988) that molecular oxygen is capable of forming superoxide from chelated iron and $NA\dot{D}$.

$$Fe^{2+}\ (EDTA) + O_2 \rightleftharpoons Fe^{3+}\ (EDTA) + \dot{O}_2^-$$

$$NA\dot{D} + O_2 \rightarrow NAD^+ + \dot{O}_2^-$$

Indeed, a Fenton reaction with Fe-EDA, H_2O_2, and NADH was effectively quenched by SOD in the presence of O_2, but not under anaerobic conditions (Imlay and Linn 1988). Of course, we do not know the equivalent of EDTA in vivo, but it is to be noted that if $\dot{O}_2^-$ is not destroyed, it will accumulate, and in the reverse of the preceding reactions, it will reduce ferric iron or NAD^+ to generate Fe^{2+} or $NA\dot{D}$, respectively.

As a final note, when superoxide is reacted on by SOD, it forms H_2O_2 and O_2. The oxygen can quench more $NA\dot{D}$ or Fe^{2+}, whereas the H_2O_2 is destroyed by catalase

$$2H_2O_2 \rightarrow 2H_2O + O_2$$

or a peroxidase

$$H_2O_2 + RH_2 \rightarrow 2H_2O + R$$

The electrons for the peroxidase reaction ultimately derive from endogenous glutathione, NADH, or NADPH.

RESPONSES TO OXIDATIVE STRESS IN *ESCHERICIA COLI*

As discussed previously, our observations led us to conclude that H_2O_2 toxicity is mediated by NADH-driven iron Fenton reactions that lead to free radicals that damage DNA. Constitutive defenses against oxygen radical toxicity support this model. They include NADH dehydrogenase, various free radical scavengers, and DNA repair proteins specific for oxidative DNA damage. As noted above previously, mutants lacking these enzyme activities are abnormally sensitive to H_2O_2 toxicity.

What about responses induced by oxidative stress? Again, most of our knowledge is for *E. coli*. In this organism, responses are remarkable both in their number and in their consistency with the preceding hypotheses.

A brief summary of the responses induced by oxygen radical stress in *E. coli* is given in Table 5-1. What is immediately apparent is that an impressive number of regulons are devoted entirely or in part to dealing with oxygen radicals. In addition, the individual responses can be divided into three categories: enhanced radical scavenging, enhanced DNA repair, or depletion of NADH and generation of NADPH.

The *oxyR* regulon responds to H_2O_2 and is devoted to the induction of enzymes that destroy free radicals. The heat-shock response overlaps the *oxyR* response partially—but not entirely. It is possibly invoked to allow the assembly of protein complexes that repair cell damage. The SOS response is brought about by oxyradical damage to DNA. Of the roughly 20 genes induced by this response, only the enhancement of RecA protein protects the cell from killing, presumably through an increased capacity for recombinational DNA repair (Imlay and Linn 1987a).

KatF is a sigma factor for RNA polymerase that appears during the entry of *E. coli* cultures into stationary phase or growth on tricarboxylic acid cycle intermediates during which free radicals are abundant. This response results in increased levels of catalase and exonuclease III, the latter enzyme being important for excision DNA repair of oxyradical damage (see Figure 5-3). In addition, NAD metabolism appears to be affected in a manner that results in a severe drop in NAD levels in the cell (Olivera, T., and Roth, J., unpublished data).

The final regulon, *soxRS*, is induced by superoxide. Among the enzymes induced are endonuclease IV, another important enzyme for excision repair of oxyradical DNA damage, and a diaphorase (NADH dehydrogenase). Induction of the latter enzyme would presumably result in decreased levels of NADH. It would possibly also act to reduce redox cycling chemicals such as paraquat so as to eliminate the production of free radicals. The final enzyme, glucose 6-phosphate dehydrogenase, catalyzes the oxidation of glucose 6-phosphate with the concomitant formation of NADPH:

$$\text{glu-6-P} + \text{NADP}^+ \rightarrow \text{6-phosphoglucono-}\delta\text{-lactone} + \text{NADPH} + \text{H}^+.$$

The significance of this induction is discussed below.

The invoking of stress responses to oxygen radicals results in enhanced DNA repair of precisely the specificity needed for DNA damage induced by oxyradicals. It also results in an enhanced ability to scavenge oxyradicals. Finally, it results in a conversion of NADH to NAD^+ (by diaphorase), then a depletion of NAD^+. (This latter depletion of NAD^+ occurs in mammalian cells by a different mechanism: DNA strand breaks result in the activation of poly [ADP ribose] polymerase, which in turn polymerizes

Table 5–1 Responses to Oxidative Stress in *Escherichia Coli*

OxyR regulon in response to H_2O_2[a,b]
- Induction of
  - Catalase (Hydroperoxidase I, *katG*)
  - Peroxidase(s)
  - Glutathione reductase
  - NADPH-dependent alkylhydroperoxidase
  - Other scavengers

Heat-shock response[a,b,c]
- Overlaps the *oxyR* response because:
  - ApppppA acts as "alaroms" for both.
  - OxyR[c] mutants are heat resistant and overexpress 9 proteins, 3 of which are heat-shock proteins.
  - H_2O_2 induced 30 proteins, 5 of which are heat-shock proteins, including the 69-kDa DnaK protein.

SOS regulon in response to DNA damage[d,e]
- Of some 20 genes induced in this response to DNA damage, apparently only increased levels of RecA protein are required for resistance to H_2O_2, presumably by enhancing recombinational DNA repair.

KatF regulon in response to oxygen stress[f,g]
- Induction of
  - Exonuclease III (*XthA*)
  - Catalase (Hydroperoxidase II, *KatE*)
  - NAD depletion[h]

SoxRS regulon in response to $\dot{O}_2^-$[i,j,k,l]
- Endonuclease IV (*nfo*)
- NADH dehydrogenase
- Glucose 6-phosphate dehydrogenase (*zwf*)

[a]Christman et al. 1985.
[b]Morgan et al. 1986.
[c]Lee et al. 1983.
[d]Imlay and Linn 1986.
[e]Imlay and Linn 1987.
[f]Sak et al. 1989.
[g]Mulvey and Loewen 1989.
[h]Olivera and Roth, unpublished data.
[i]Greenberg and Demple 1989.
[j]Kogoma et al. 1988.
[k]Weiss et al. 1990.
[l]Chan and Weiss 1987.

the NAD^+ to poly[ADP ribose] [Imlay and Linn, 1988].) These responses would be those predicted from the hypotheses shown in Figure 5-4.

THE ROLES OF NADPH IN PROTECTING THE CELL AGAINST OXYRADICALS

The increase of NADPH levels as a result of the induction of glucose 6-phosphate dehydrogenase during the *oxyRS* response can be rationalized by the facts that NADPH is the cofactor for glutathione reductase and that both NADPH and reduced glutathione are cofactors for peroxidases. Glutathione is also important for maintaining proteins and iron in the reduced state—certainly in red blood cells. However, if the cell depletes NADH so as to prevent the driving of Fenton reactions, cannot NADPH assume that role? The answer is no—NADPH is not readily oxidized by H_2O_2 and iron, and it does not drive a DNA-nicking reaction (Chin et al., unpublished data). Indeed, NADPH inhibits DNA nicking in the reaction of $Fe^{3+}/NADH/H_2O_2$ (Chin, S., Imlay, J. A., and Linn, S., unpublished data). In essence, NADPH is able to sequester free iron away from NADH and into a nonredox state. Hence, NADPH serves two important roles in protecting the cell: It provides or acts as a cofactor for the reduction of radicals, and it removes iron from Fenton redox reactions.

FUTURE PERSPECTIVES

As discussed elsewhere in this volume, oxygen radicals and the damage they cause are now implicated in arthritis, heart disease, cancer, and degenerative phenomena during the aging process. Moreover, as discussed in this chapter and elsewhere in the volume, it appears that DNA damage is a major target of oxyradicals and that such damage could be important in these disease states. Moreover, iron plays a significant role in the processes that cause this damage, both in procaryotes and eucaryotes. Hence, it is likely that molecular biologists, cell biologists, biochemists, and pharmacologists will turn toward monitoring DNA damage caused by oxygen radicals as a dosimeter of oxyradical damage. Damage to DNA caused by oxygen radicals will be better defined when reliable assays for such damage are developed. We ought also to turn our attention to iron-binding agents (in addition to free-radical scavengers) as antagonists of the generation of DNA damage and other forms of cellular damage. Finally, efforts must increase toward understanding how DNA damage is repaired, and whether, and to what extent, deficiencies in these repair processes might lead to predispositions to the various diseases with which free radical damage has been implicated.

ACKNOWLEDGMENTS

Research cited from our laboratory was supported by grant R29 GM19020 from the NIH. The author thanks Bruce Demple, James Imlay, and S. M. Chin, who not only did much of the work cited in this review, but also provided much of the intellectual stimulus, and Drs. Spencer Farr, Bruce Demple, John Roth, and Baldomera Olivera for discussing, and allowing him to cite, their unpublished results.

REFERENCES

Brimer, L. H., and Lindahl, T. (1984). Excision of oxidized thymine from DNA. *J. Biol. Chem.*, 259:5543–48.

Chan, E., and Weiss, B. (1987). Endonuclease IV of *Escherichia coli* is induced by paraquat. *Proc. Natl. Acad. Sci. USA*, 84:3189–93.

Christman, M. F., Morgan, R. W., Jacobson, F. S., and Ames, B. N. (1985). Positive control of a regulon for defenses against oxidative stress and some heat shock proteins in *Salmonella typhimurium. Cell*, 41:753–62.

Franklin, W. A., and Lindahl, T. (1988). DNA deoxyribosphosphodiesterase. *EMBO J.*, 7:3617–21.

Greenberg, J. T., and Demple, B. (1989). A global response induced in *Escherichia coli* by redox-cycling agents overlaps with that induced by peroxide stress. *J. Bacteriol.*, 171:3933–39.

Gutman, M., and Eisenbach, M. (1973). On the complexation of ferric ions by reduced nicotinamide adenine dinucleotide. *Biochemistry*, 12:2314–17.

Gutman, M., Margalit, R., and Schejter, A. (1968). A charge-transfer intermediate in the mechanism of reduced diphosphopyridine nucleotide oxidation by ferric ions. *Biochemistry*, 7:2778–85.

Imlay, J. A., and Linn, S. (1986). Bimodal pattern of killing of DNA-repair-defective or anoxically grown *Escherichia coli* by hydrogen peroxide. *J. Bacteriol.*, 166:519–27.

Imlay, J. A., and Linn, S. (1987a). Mutagenesis and stress responses induced in *Escherichia coli* by hydrogen peroxide. *J. Bacteriol.*, 169:2967–76.

Imlay, J. A., and Linn, S. (1987). Toxicity, mutagenesis and stress responses induced in *Escherichia coli* by hydrogen peroxide. *J. Cell Sci. Suppl.*, 6:289–301.

Imlay, J. A., and Linn, S. (1988). DNA damage and oxygen radical toxicity. *Science*, 240:1302–09.

Imlay, J. A., Chin, S. M., and Linn, S. (1988). Toxic DNA damage by hydrogen peroxide through the Fenton reaction *in vivo* and *in vitro*. *Science*, 240:640–42.

Johnson, A. W., and Demple, B. (1988). Yeast DNA diesterase for 3′-fragments of deoxyribose: Purification and physical properties of a repair enzyme for oxidative DNA damage. *J. Biol. Chem.*, 263:18009–16.

Kim, J., and Linn, S. (1988). The mechanism of action of *E. coli* endonuclease III and T4 UV endonuclease (endonuclease V) at AP sites. *Nucleic Acids Res.*, 16:1135–41.

Kogoma, T., Farr, S. B., Joyce, K. M., and Natvig, D. O. (1988). Isolation of gene fusions (*soi::lacZ*) inducible by oxidative stress in *Escherichia coli. Proc. Natl. Acad. Sci. USA*, 85:4799–4803.

Lee, P. C., Bochner, B. R., and Ames, B. N. (1983). AppppA, heat-shock stress, and cell oxidation. *Proc. Natl. Acad. Sci. USA*, 80:7496–7500.

Morgan, R. W., Christman, M. F., Jacobson, F. S., Storz, G., and Ames, B. N. (1986). Hydrogen peroxide-inducible proteins in *Salmonella typhimurium* overlap with heat shock and other stress proteins. *Proc. Natl. Acad. Sci. USA*, 83:8059–63.

Mulvey, M. R., and Loewen, P. C. (1989). Nucleotide sequence of *katF of Escherichia coli* suggests katF protein is a novel σ transcription factor. *Nucleic Acids Res.*, 17:9979–91.

Sak, B. D., Eisenstark, A., and Touati, D. (1989). Exonuclease III and the catalase hydroperoxidase II in *Escherichia coli* are both regulated by the *katF* gene product. *Proc. Natl. Acad. Sci. USA*, 86:3271–75.

Spiro, S., Roberts, R. E., and Guest, J. R. (1989). FNR-dependent repression of the *ndh* gene of *Escherichia coli* and metal ion requirement for FNR-regulated gene expression. *Mol. Microbiol.*, 3:601–08.

Wallace, S. S. (1988). AP endonucleases and DNA glycosylases that recognize oxidative DNA damage. *Environ. Mol. Mutagen.* 12, 432–77.

Weiss, B., Tsaneva, I. R., and Wu, J. (1990). A superoxide response resulon in *Escherichia coli.* In: S. Wallace, and R. Painter eds. *UCLA Symposia on Molecular and Cellular Biology, New Series*, New York: Wiley Liss, 287–95.

6

Gene Regulation by Active Oxygen and Other Stress Inducers: Role in Tumor Promotion and Progression

NANCY H. COLBURN

Many of the cellular responses elicited by reactive oxygen also occur in response to ultraviolet (UV) or ionizing radiation or to treatment by cytokines, growth factors, hormones, or tumor promoters. In *Escherichia coli*, DNA damage caused by UV radiation or certain chemicals induces some 20 genes as part of the SOS response (Chapter 5). Some of these *E. coli* genes are also inducible by heat or oxidant stress. Certain nonoxidant stress inducers produce in eukaryotic cells elevation of active oxygen, and in some cases there is evidence that the resultant oxidative stress is involved in producing a biologic end point such as altered growth or differentiation or neoplastic transformation. Recent advances in understanding the mechanisms of such stress-induced responses make possible an inquiry into whether two or more such stress inducers act on convergent pathways to produce proliferative stimulation or differentiation or cancer. This chapter focuses on such an inquiry. In particular we address the following questions:

1. What are the genes whose expression is altered by exposure of cells to generators of superoxide anion, H_2O_2, or hydroxyl radicals, or to tumor necrosis factor, x-ray or UV irradiation, or to the tumor-promoting phorbol ester 12-*O*-tetradecanoylphorbol-13-acetate (TPA)?
2. To what extent do these treatments produce common signal transduction or gene expression events? Are there common modes of transcriptional or post-transcriptional regulation?
3. To what extent are these stress-induced gene expression events causally related to tumor induction or tumor progression?

Table 6-1 summarizes some 21 genes or classes of genes whose expression

Table 6–1 Gene Expression in Response to Oxidant or Other Stress[a]

Gene	Inducer						
	X/XO	H_2O_2	Fe	TNF-α	X-ray	UV	TPA
c-*myc*	↑			↑ or =			↑[b]
c-*fos*	↑	↑[b]		↑ or =		↑[b]	↑[b]
c-*jun*			↑			↑[c]	↑[b]
Collagenase I			↑[b,c]			↑	↑[b]
Cu/Zn SOD					=		↓[b]
Catalase							↓[b]
MnSOD				↑	↑	↑	
IL-1,6				↑			
Collagen I				↓[c]			↓[b]
SV40 enhancer						↑	↑
HIV-1-CAT						↑	↑[b]
MoLTR-CAT					↑[b]	↑[b]	↑[b]
Metallothionine I, II	=				=	↑	↑
Heme oxygenase	↑					↑	
Class I DDI	=	=			=	↑	=
Class II DDI	↑	↑			=	↑	=
β DNA polymerase	↑				=		=
EPIF	↑					↑	
Nmo-1	↑					↑	
ODC						↑	↑
PA						↑	↑
TFR			↓				
Ferritin			↑				

[a](↑) = mRNA levels are increased; (↓) = mRNA levels are decreased (↓); (=) = mRNA levels are unchanged; DDI = DNA damage-inducible; EPIF = extracellular protein synthesis-inducing factor; ODC = ornithine decarboxylase; PA = plasminogen activator; TFR = transferrin receptor; SOD = superoxide dismutase; IL = interleukin; HIV = human immunodeficiency virus; CAT = chloramphenicol acetyl transferase; MoLTR = Moloney viral long-terminal repeat; X/XO = xanthine/xanthine oxidase; TNF-α = tumor necrosis factor α/γ; UV = ultraviolet light; TPA = 12-0-tetradecanoyl phorbal-13-acetate; EPIF = extracellular protein synthesis-inducing factor; PA = plasminogen activator.

[b]PKC dependent.

[c]Cycloheximide blocks.

is induced or repressed by oxidant or other stress inducers in a variety of cells in culture. At present there is little or no evidence for cell type specificity of the gene expression responses listed. Table 6-2 lists signal transduction events occurring in response to these stress inducers that may lead to altered gene expression. What follows is organized according to stress inducer.

XANTHINE/XANTHINE OXIDASE, H_2O_2, OR IRON-INDUCED GENE EXPRESSION

Active oxygen species, particularly hydroxyl radicals, produce both single- and double-strand breaks in DNA (Mellow Filho et al. 1984). Whether this DNA damage occurs by a direct or indirect mechanism has not been established, but there is evidence favoring the involvement of radical-induced changes in Ca^{2+} flux and the activation of Ca^{++}-dependent nucleases (Cantoni et al. 1989). Cantoni and colleagues (1989) found that the calcium chelator Quin 2 prevents H_2O_2-induced DNA breaks and cytotoxicity, suggesting that H_2O_2 produces DNA damage by altering intracellular calcium signaling. An alternative interpretation is that cellular peroxidases catalyze a reaction between Quin 2 and H_2O_2 that depletes the H_2O_2. Oxygen radicals may also act to activate protein kinases such as calcium/calmodulin-dependent, ribosomal S6, and calcium/phospholipid-dependent kinases (PKC) (Larsson and Cerutti 1989; Gopalakrishna and Anderson 1989), and altered gene regulation may occur as a consequence of such altered kinase activity. Gopalakrishna and Anderson (1989) found that, although H_2O_2 can oxidatively inactivate purified PKC, it can also, under milder conditions, produce selective oxidative modification of the regulatory domain that appears to negate the requirement for Ca^{++} and phospholipids for activation of PKC. Both DNA damage and activation of Ca^{++} calmodulin-dependent and S6 kinases occurred within 15 to 30 min of exposure of JB6 mouse epidermal cells to xanthine/xanthine oxidase (x/xo) (Krupitza and Cerutti 1989; Larsson and Cerutti 1988; Muehlematter et al. 1988). Kinase activation was blocked by Quin 2, indicating its dependence on intracellular free calcium (Larsson and Cerutti 1988). Phosphorylation of the PKC substrate adenosine diphosphoribosyl transferase (ADPRT) and of the S6 kinase substrate, the ribosomal S6 protein (small ribosomal subunit protein number 6), occurred within 15 to 20 minutes (Larsson and Cerutti 1988, 1989). Based on differential sensitivity to catalase and Cu^{++}/Zn^{++} superoxide dismutase (SOD), the ADPRT phosphorylation following X/XO appears to depend on the elevation of $\dot{O}_2^-$ rather than that of H_2O_2, whereas S6 phosphorylation depends on elevation of H_2O_2, not $\dot{O}_2^-$ (Larsson and Cerutti 1988, 1989). S6 phosphorylation is PKC independent as indicated by its occurrence following PKC down-regulation (Larsson and Cerutti 1989). ADPRT phos-

Table 6–2 Signal Transduction in Response to Oxidant or Other Stress[a]

Signal	Inducer					
	$X/XO \rightarrow \dot{O}_2^-$	H_2O_2 or $X/XO \rightarrow\rightarrow H_2O_2$	TNF-α	X-ray	UV	TPA
DNA damage	↑	↑	↑	↑	↑	=?
PKC activation	↑?	↑ & ↓	↑	↑	↑	↑[b]
Ca^{++} ↑		↑[c]			↑[b]	
Ca^{++}/calmodulin PK activation		↑[c]				
S6 Kinase activation		↑[c]				
Phos of ADPRT	↑[b]					
polyADPR of Topoisomerase I	↑[b]					
polyADPR of ADPRT	↑[b]					

[a](↑) = signal is activated or increased; (↑) = signal is inactivated or decreased; (=) = signal is unchanged; PKC = calcium/phospholipid-dependent kinases; ADPRT = adenosine diphosphoribosyl transferase; X/XO = xanthine/xanthine oxidase; TNF-α = tumor necrosis factor α; UV = ultraviolet; TPA = 12-0-tetradecanoyl phorbal-13-acetate.

[b]PKC-dependent.

[c]PKC-independent.

phorylation in vivo and histone HI phosphorylation in vitro were stimulated by the sulfhydryl reagent diamide, suggesting that PKC is activated by oxidation (Larsson and Cerutti 1988). Polyadenosine diphosphate (poly ADP) ribosylation of chromosomal proteins ADPRT and topoisomerase I occurs in 1 h or more in JB6 mouse epidermal cells in response to X/XO-induced DNA damage and inactivates the corresponding enzymatic activities (Krupitza and Cerutti 1989; Muehlematter et al. 1988). Finally, transient elevation of c-*fos* mRNA within 1 h (10-fold) and c-*myc* mRNA within 4 or more h (5-fold) of X/XO occurs (Crawford et al. 1988). Elevated expression of proto-oncogenes c-*fos*, c-*jun*, or c-*myc* is known to occur in response to a number of mitogenic and nonmitogenic stimuli as well as constitutively in various tumors, but whether elevated expression is causally related to induction of neoplastic transformation is at present unclear. Figure 6-1 summarizes what is currently known of oncogenes, their intracellular localization, and their functions. The X/XO-induced events that trigger the elevations of *fos* and *myc* mRNAs in JB6 cells are not known, but it is known that the c-*fos*, but not c-*myc*, mRNA elevation involves transcriptional activation as shown by nuclear runoff assays (Crawford et al. 1988).

In addition to *fos* and *myc*, other active oxygen-inducible genes have been described. One is heme oxygenase (Keyse and Tyrrell 1989), also induced by UV light and discussed below. Others are subsets of UV-inducible genes described by Fornace and coworkers (1988) that are also inducible by H_2O_2, by the alkylating agent methyl methane sulfonate, or by growth arrest but not by heat shock or by phorbol esters. These are described further in the section on UV-inducible gene expression.

Because cellular iron catalyzes the generation of hydroxyl radicals and because iron metabolism may be important in cancer etiology (see Chapter 7), the subject of iron-induced gene regulation is considered here. Unlike the genes discussed in this chapter whose regulation involves protein binding to DNA elements, the two genes that mediate the uptake and detoxification of iron are regulated post-transcriptionally by protein-RNA binding (Klausner and Hartford, 1989). Iron uptake is mediated by diferric transferrin bound to the transferrin receptor (TFR). In the cytoplasm, iron is used for iron-requiring processes or sequestered in ferritin. When iron is plentiful, the number of TFRs is decreased and the level of ferritin is increased, whereas limiting iron produces the opposite result. These adjustments arise as a consequence of the binding of a particular protein (BP) to an iron response element (IRE) on the respective mRNAs (Klausner and Harford 1989). When the BP binds to the IRE of TFR mRNA at the 3′ end, it protects against mRNA degradation; but high-affinity binding of the IRE-BP occurs when iron is *not* bound, thus explaining elevated TFR mRNA stability and receptor synthesis when iron is limiting. When the BP binds to the IRE of ferritin mRNA at the 5′ end, it blocks ferritin translation, but high-affinity BP binding and blocked translation occur

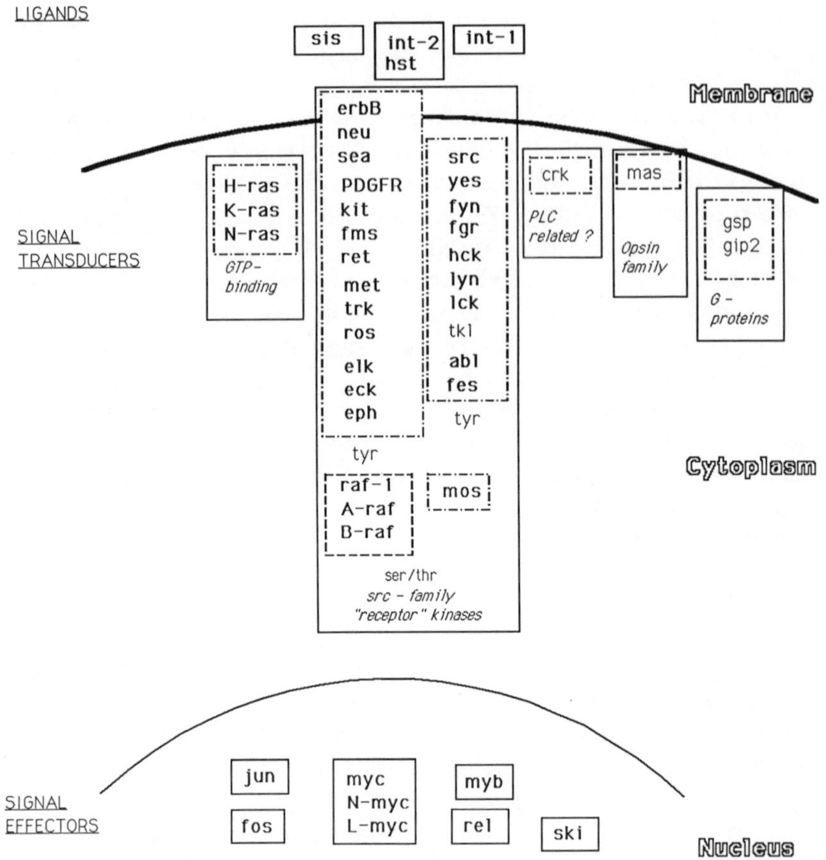

Figure 6–1 Summary of known localizations and functions of proto-oncogene products. The classes of oncogene products shown include growth factor ligands for cell surface receptors; plasma membrane receptor tyrosine kinases; membrane-associated tyrosine kinases; GTP-binding "G" proteins; other plasma membrane proteins; cytoplasmic serine/threonine kinases; and nuclear oncogene proteins, several of which are known to bind to DNA and to regulate gene expression. The dashed boxes enclose genes showing significant sequence homology. The solid boxes enclose evolutionarily related genes. Genes in bold type have been shown to have oncogenic activity whereas the others have not. (Reprinted with permission from Kolch et al. 1991.)

when iron is *not* bound, thus explaining elevated ferritin synthesis when iron is plentiful.

TFN-α-INDUCED GENE EXPRESSION

Tumor necrosis factor α (TNF-α) is a multifunctional cytokine secreted by macrophages in response to inflammation, infection, and cancer (Beutler and Cerami 1989). TNF-α binds to high-affinity cell surface receptors (Aggarwal et al. 1985; Beutler and Cerami 1989) to trigger signal transduction events that regulate gene expression and eventually to bring about protease or cytokine secretion or cytotoxicity. The nature of the signal transduction events is not fully understood, but TNF-α appears to cause increased production of $\dot{O}_2^-$ from mitochondria (Boveris 1977; Wong et al. 1989). The elevated $\dot{O}_2^-$ or derivative radicals putatively produce DNA damage. TNF-α also activates a cellular DNase that may be responsible for DNA damage (Duke et al. 1983; Smith et al. 1990). In TNF-α-treated human fibroblasts, elevation of *fos* and *myc* mRNAs occurs early, as does elevation of *jun* mRNA (Brenner et al. 1989). Following the elevation of *jun* and *fos* mRNAs, AP-1 (Jun/Fos complex)-mediated transcription of collagenase is activated at 6 to 24 hrs (Brenner et al. 1989). This response to TNF-α appears to be PKC dependent and to involve new-protein synthesis, as indicated by its sensitivity to the PKC inhibitor H-7 and to the protein synthesis inhibitor cycloheximide. This transcriptional activation response to TNF-α is not accompanied by mitogenic stimulation.

Another gene whose expression is induced by TNF-α is the mitochondrial manganese superoxide dismutase (MnSOD) (Wong et al. 1989). Whether MnSOD induction is dependent on PKC or on any other kinases is unknown. In many tumor cells, in contrast to normal cells, this enzyme is not inducible by TNF, hence possibly explaining the selective toxicity of TNF-α to tumor cells (Wong et al. 1989; Oberley and Oberley 1988). TNF-sensitive cervical carcinoma cells can be made TNF-resistant by introduction of an MnSOD expression construct, but not by introduction of an antisense MnSOD (Wong et al. 1989), thus providing support for the hypothesis that $\dot{O}_2^-$ mediates the cytotoxic response to TNF-α. Other antioxidant enzymes that appear not to be involved in protection against TNF toxicity include Cu/Zn SOD, glutathione peroxidase, and catalase, as demonstrated by the observation that the transfectant cell lines that acquired TNF resistance by overexpressing MnSOD showed no greater expression of these other enzymes than that seen in the parental TNF-sensitive lines (Wong et al. 1989).

The antimitogenic action of TNF-α is also associated with the induction of c-*jun*. TNF inhibition of basic fibroblast growth factor (bFGF)-induced vascular endothelial cell proliferation was accompanied by induction of c-*jun* but not of c-*myc* or c-*fos* transcription (Dixit et al. 1989). Unlike the cytotoxic effects protected by induced MnSOD described previously, the antiproliferative effects associated with c-*jun* induction were reversible, suggesting that the targets of $\dot{O}_2^-$ differ in the two situations. TNF evokes expression of immediate early genes such as *jun* independently of protein synthesis (Dixit et al. 1989) as do growth factors and other inducers discussed here. In contrast to observations with growth factor-treated human or rodent fibroblasts, the preceding results with vascular endothelial cells illustrate that *jun* is not necessarily regulated in coordination with *fos* and *myc* expression and that *jun* induction is not necessarily associated with mitogenic stimulation. Finally, TNF-α inhibits human fibroblast collagen

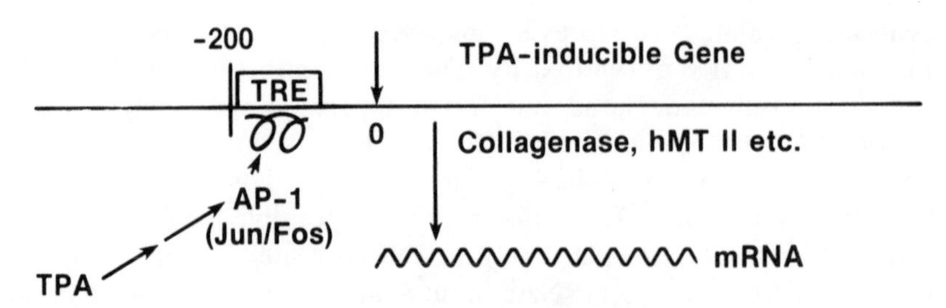

Figure 6–2 Transcriptional transactivation triggered by a tumor promoter. Tumor-promoting phorbolesters or certain growth factors (58) stimulate the formation of AP-1 (*Jun/Fos*) heterodimers that bind to the so-called TPA response element (TRE) (cis element whose sequence is ATGAGTCAG) usually located within 200 nucleotides upstream of the transcriptional start of a gene. Consequent to the binding of the AP-1 transacting factor, the transcription of the attached gene occurs. AP-1 binding to the TRE is known to activate the transcription of a number of genes including collagenase and metallothionein (Bernstein and Colburn 1989).

type I transcription by a mechanism dependent on new protein synthesis without cytotoxicity (Solis-Herruzo et al. 1988).

INDUCTION OF GENE EXPRESSION BY ULTRAVIOLET OR X-RAY EXPOSURE

UV light is absorbed by DNA with the resultant production of photoproducts such as thymine dimers and DNA strand breaks. The action spectrum for UV-induced gene expression corresponds to the absorption spectrum for DNA damage and cell killing (Stein et al. 1989). Stein and coworkers (1989) analyzed UV-induced gene transcription by introducing, into HeLa cells, bacterial chloramphenicol acetyltransferase (CAT) constructs driven by one of three different transcriptional promoters harboring known response elements. Figure 6-2 provides a summary of transcriptional transactivation of gene expression involving the binding of specific proteins (transactivating factors) to specific DNA sequences (cis elements) in a gene's promoter near its transcriptional start. The three transcriptional promoters studied by Stein and colleagues (1989) included the human immunodeficiency virus-1 (HIV-1) promoter that contains a cis element activated by the binding of the protein NFκB (Nabel and Baltimore 1987), the collagenase I promoter activated by the binding of AP-1 (Jun/Fos) complexes to a 9-nucleotide cis element (Angel et al. 1987), and the *fos* promoter activated by the binding of 67 and 62 kDa serum response proteins (SRPs) to a serum response element (Triesman 1985). Because each of these promoters was driving a CAT reporter in a recombinant DNA construct, activation of the corresponding promoter following UV treatment of the transfected HeLa cells could be assessed by measuring induced CAT enzyme activity (reflecting CAT transcription and translation). Thus UV radiation was found to activate the NFκB-dependent HIV promoter, the Jun/Fos-dependent collagenase promoter, and the SRP-dependent *fos* promoter (Stein et al. 1989).

These responses appear to occur independently of UV-induced DNA repair, because the response in xeroderma pigmentosum UV-repair-deficient cells was undiminished (Stein et al. 1989). Induction of these mRNAs occurs within minutes and, for factors binding to the HIV-1 and *fos* promoters, initially involves post-translational modification of these DNA-binding proteins (Stein et al. 1989). UV induction of *fos* is protein-synthesis independent (Stein et al. 1989), as indicated by its cycloheximide insensitivity, and PKC dependent (Buscher et al. 1988), as suggested by H-7 inhibition. UV causes secretion of an extracellular protein synthesis-inducing factor (EPIF) (Schorpp 1984) from a variety of cell types that activates the NFκB-dependent transcription from the HIV promoter (Stein et al. 1990) and also activates collagenase transcription (Whitman et al. 1986), thus suggesting a mechanism of spreading UV effects to human cells in vivo that are not exposed to UV radiation. It is noteworthy that NFκB is known to be activated by release from an inhibitor in the cytoplasm, thus indicating that UV-induced signal transduction can pass through the cytoplasm (Nabel and Baltimore 1987). Others have reported a slower (than minutes) HIV transcriptional induction with a peak at 20 h post UV irradiation (Valerie et al. 1988). An 18-h peak has been seen for transcription promoted by Moloney viral long-terminal repeat (MoLTR) (Lin et al. 1990). This UV effect is dependent on PKC but is independent of AP-1 or AP-2 transacting factors, because the cis elements to which AP-1 or AP-2 proteins bind are absent from the MoLTR (Lin et al. 1990). It has been noted by Stein and colleagues (1990) that the UV doses found to activate a virus such as HIV-1 are in the range calculated to occur during 30 to 40 min of noontime sun exposure in Texas or the Rocky Mountains.

X-ray irradiation produces DNA damage that differs from that induced by UV radiation and is repaired differently but, like UV, produces changes in gene expression. X-ray irradiation induces or enhances transcription from the murine sarcoma virus LTR, but with a shorter time course than that seen after UV exposure (Lin et al. 1990). This induction is PKC dependent, but AP-1 and AP-2 independent. Lin and colleagues (1990) suggested that the time courses for UV and x-ray regulation may be related to the respective half-lives for DNA repair (Summers et al. 1989), but the above-described data with xerodema pigmentosum (Stein et al. 1989) would argue against this possibility for UV effects. X-ray irradiation also induces MnSOD in mouse heart (1 h) (Oberley et al. 1987) and in intestinal muscle (20 h) (Valerie et al. 1988).

UV treatment of cultured human skin fibroblasts induces heme oxygenase (32 kDa) (Keyse and Tyrrell 1989) at 2 h, which then enhances bilirubin production. Bilirubin in turn provides protection against UV irradiation (Stocker et al. 1987). Fornace and coworkers (1988) have isolated some 49 DNA damage-inducible (DDI) cDNAs after subtraction hybridization of control mRNA from UV-irradiated Chinese hamster cell (CHO or V79) cDNA. Except for two metallothionein genes and part of one other gene, these had sequences unlike others in the GenBank. The genes of one DDI class were inducible by UV, but not by the alkylating agent methylmethanesulfonate (MMS), by H_2O_2, by heat shock, by TPA or by x-irradiation. A second class of DDI genes were inducible by UV irradiation and MMS. A subset of five of these were inducible also by H_2O_2, and most of these H_2O_2- and UV-inducible genes were also inducible by prolonged growth inhibition (Fornace et al. 1989). Most of these DDI genes were not inducible by TPA or heat shock, thus illustrating the generalization that, in contrast to those in bacteria (see Chapter 5), mammalian heat shock response genes are not necessarily genes that respond to a broad range of stresses including UV. (Papathanasiou and Fornace 1991).

Those c^{14Cos}/c^{14Cos} mice harboring a homozygous deletion on chromosome 7 for NAD(P)H menadione oxidoreductase (NMOI), a member of the Ah receptor gene battery and an enzyme thought to protect against oxidative stress, were found to express three of the above five DDI genes at a markedly elevated level, compared with wild-type mice (Gluecksohn-Waelsch 1987; Fornace et al. 1989). Overexpression of these genes may be due to the loss of a specific negative regulatory factor encoded by a gene located in the deleted portion of chromosome 7. This may be analogous to the *E. coli* SOS response mediated by removal of the lexA repressor (see Chapter 5).

TUMOR PROMOTERS, HORMONES, AND GROWTH FACTORS

Phorbol esters and certain growth factors produce many of the same gene expression responses seen with active oxygen and radiation. These include: induction of proto-oncogenes c-*fos*, c-*jun*, and c-*myc*, and proteases such as collagenase via transactivators such as AP-1 in the case of collagenase or SRE in the case of *fos* activation (Buscher 1988; Stein et al. 1989). Also responsive to TPA are HIV (Stein et al. 1989, 1990) and MoLTR (Lin et al. 1990) retroviral promoter-driven gene expressions that involve NFκB and other non-AP-1 binding factors, respectively, as well as transcriptional induction of UV-inducible ornithine decarboxylase (ODC) (Verma et al. 1988), plasminogen activator (Miskin and Beni-Ishai 1981), and heme oxygenase (Kageyama et al. 1988). TPA treatment of mouse JB6 cells decreases mRNA levels for Cu^{++}/Zn^{++} SOD and catalase by a not yet elucidated mechanism that involves no decrease in transcription (Crawford et al. 1989).

Estrogens constitute another class of compounds that act as tumor promoters or in some cases as complete carcinogens. Diethylstilbestrol (DES) is known to form superoxide radicals and to induce DNA strand breaks (Epe et al. 1986) by a mechanism that appears to involve DES-semiquinone formation. Estrogen induces oxidative stress-associated enzymes (Roy and

Liehr 1989) and ODC (Sheehan and Branham, 1987), as well as a 24-kDa heat shock protein (Fuqua et al. 1989).

RELATIONSHIP OF STRESS-INDUCED GENE EXPRESSION TO NEOPLASTIC TRANSFORMATION

Many of the stress inducers discussed previously are mitogenic. TNF-α usually is not. All of them can promote or induce neoplastic transformation in some systems. Although mitogenic stimulation may be required as one component of tumor promotion, it appears not to be sufficient alone (Colburn et al. 1981). Changes in gene expression that produce escape from growth regulation or altered differentiation or other altered responses appear to be necessary events. Whether certain of the above-described active oxygen-induced genes are transformation effector genes is the subject of active investigation. Mouse epidermal JB6 genetic variant cell lines are differentially responsive to induction of neoplastic transformation by phorbol esters, growth factors, and other tumor promoters (Garrity et al. 1989). Promotion sensitive (P^+) JB6 variants can be transformed to anchorage independence and tumorigenicity by phorbol esters, growth factors, and xanthine (xanthine oxidase), whereas promotion-insensitive (P^-) JB6 cells are resistant to such induction of neoplastic transformation. Studies by Larsson and Cerutti (1988, 1989) have demonstrated differential DNA damage and associated polyADP ribosylation in JB6 promotion-sensitive and promotion-resistant cells. Following X/XO treatment, P^+ cells showed less DNA damage and less polyADP ribosylation (Muehlematter et al. 1988; Krupitza and Cerutti 1989), as well as less induction of c-*myc* and c-*fos* mRNA accumulation (Crawford et al. 1988) compared with the resistant P^- cells. Cerutti (1989) has proposed that promotable (P^+) cells possess a response modification that manifests as a superior antioxidant defense that protects them from the excessive cytostatic effects of active oxygen. An alternative interpretation is that c-*myc* and c-*fos* induction may be a consequence of DNA damage and polyADP ribosylation and that induced expression of c-*myc* and c-*fos* is not limiting for the transformation response to X/XO and may not even be necessary. It must be noted that direct evidence for the P^+/P^-differential transformation responsiveness to X/XO-produced superoxide anion has not been reported. Indirect evidence comes from the finding that an elevated level of superoxide anion is an essential early mediator of TPA-induced transformation that occurs in JB6 P^+ but not P^- cells (Nakamura et al. 1985, 1988). Other approaches to studying the role of antioxidant defense in neoplastic transformation have been reported by Seidegard and coworkers (1986, 1988). These authors have implicated the antioxidant defense enzyme glutathione transferase (GST) in lung cancer susceptibility. Patients with hereditary deficiencies in the expression of leukocyte glutathione transferase, due to a gene deletion, show substantially increased susceptibility to lung cancer. Oltipraz, a substituted 1,2-dithiole-3-thione that inhibits the rate of transcription and levels of GST mRNA (Davidson et al. 1990), also inhibits carcinogenesis induced by benzopyrene, diethylnitrosamine, or uracil mustard in mouse lung or forestomach (Wattenberg and Bueding 1986) or by aflatoxin B1 in rat liver, by enhancing detoxication or by blocking carcinogen activation or both (Kensler et al. 1987).

The data summarized in Tables 6-1 and 6-2 suggest two possibilities for common events that may mediate the transforming action of several of the stress inducers considered thus far. The first is activation of protein kinase C with or without consequent activation of another protein kinase. PKC inhibition and downmodulation experiments implicate PKC activation in transformation or gene regulation events that are triggered by exposure not only to phorbol esters but also to TNF-α, x-irradiation, and UV light. Alternatively, C kinase and other kinases might converge at a step involving phosphorylation and consequent activation of a common substrate. Evidence favoring a causal role for PKC overexpression in neoplastic transformation comes from the observation by Housey and colleagues (1988) that introduction of a PKC β expression construct into rat fibroblasts with consequent overexpression of PKC β protein is accompanied by an increase in transformed phenotype.

The second class of contenders for a common event is the group of transcriptional transactivators such as AP-1, NFκB, serum factors, and other factors known to transactivate collagenase, HIV, c-*fos*, and MoLTR-driven gene expression in response to TPA, TNF-α, UV, and x-irradiation. Transactivation involving binding of an AP-1 (Fos/Jun) complex to activate collagenase expression has been reported to occur in response to TPA, TNF-α, and UV radiation (See Table 6-1). Mouse epidermal JB6 P^+ but not P^- cells showed TPA- or EGF-induced AP-1-dependent transactivation of gene expression when a transfected AP-1 binding cis element in a TK promoter-driven CAT reporter was assayed (Bernstein and Colburn 1989). This finding suggests that AP-1 (Jun/Fos complex)-dependent gene expression may be required for promotion of transformation and that lack of it may in part explain the resistance of the P^- variants. Recent results show higher levels in P^+ cells of TPA-induced c-*jun* but not of c-*fos* gene expression (Cerutti 1989; Bernstein et al. 1991; Ben-Ari, E., and Colburn, N. H., unpublished data). Thus, unless Fos and Myc proteins are negative regulators, differential *fos* or *myc* expression and associated DNA damage and polyADP ribosylation (see Krupitza and Cerutti 1989; Muehlematter et al. 1988) appear not to explain the differential sensitivity to promotion of neoplastic transformation by X/XO or TPA in JB6 P^+ and P^- cells. In contrast, c-*jun* expression and consequent regulation of AP-1-dependent transcriptional transactivation may contribute to determining the responsiveness of JB6 P^+ cells to promotion of neoplastic transformation, at least by TPA. A careful analysis of the genes transcriptionally regulated by AP-1 may identify critical promotion-relevant effector genes. Additional support for a role of *jun* overexpression in transformation comes from the observation of Schütte and coworkers (1989) that introduction of a c-*jun* expression construct into rat-1 fibroblasts produced transformed foci and a tumorigenic phenotype.

CONCLUSIONS AND FUTURE DIRECTIONS

Some of the stress inducers discussed above produce DNA damage that in turn appears to induce changes in gene expression. Other stress inducers such as phorbol esters, growth factors, and most hormones, whether or not they produced elevated active oxygen, appear not to produce DNA damage but nevertheless produce many of the gene expression changes seen with DNA damaging agents; these gene expression events are often regulated by the same transcriptional transactivation processes. Some of the immediate questions that need to be answered concern the details of whether mRNA levels are regulated at the level of transcription rate or of mRNA stability; what are the identities of the DNA response elements and DNA binding proteins; and what are the early signal transduction events, such as kinase activation, for genes such as Cu/Zn- and MnSOD, the MoLTR-driven viral genes, heme oxygenase, ornithine decarboxylase, and the DNA damage inducible genes. Although the answers are known for genes such as collagenase and HIV genes, further understanding of these other genes may yield insight into possible common pathways for stress inducers. If there are common pathways and if they can be shown to be critical to neoplastic transformation, such events could be targets for chemoprevention.

If one wishes to show a critical role for stress-induced gene expression in neoplastic transformation, one can attempt to produce neoplastic transformation by overexpressing the gene(s) in question or attempt to block neoplastic transformation by inhibiting expression of the gene(s) in question using antisense constructs or other approaches. Evidence supporting a causal role has been forthcoming in the case of c-kinase overexpression or c-*jun* overexpression in combination with other oncogene alterations (Housey et al. 1988; Bernstein et al. 1991). Additional early-induced genes need to be examined to meet the predictions for a causal relationship to neoplastic transformation.

Finally, the question of effector genes needs to be addressed. Early-induced genes such as the proto-oncogenes *fos*, *jun*, and *myc* encode proteins that function in the nucleus as transcriptional transactivators or other regulators of gene expression. The search for the answers to the questions, what are the effector genes that are regulated by the early-induced genes, and which of these bring about neoplastic transformation, has only begun.

REFERENCES

Aggarwal, B., Eessalu, T., and Hass, P. (1985). Characterization of receptors for human tumour necrosis factor and their regulation by γ-interferon. *Nature*, 318:665–67.

Angel, P., Imagawa, M., Chiu, R., Stein, B., Imbra, R. J., Rahnsdorf, H. J. et al. (1987). Phorbol ester-inducible genes contain a common cis element recognized by a TPA-modulated trans-acting factor. *Cell*, 49:729–39.

Ben-Ari, E. T., Bernstein, L. R., Colburn, N. H. (1992). Differential C-*jun* expression in response to tumor promoters in JB6 cells sensitive to neoplastic transformation. *Mol. Carcinog.*, 5 (in press).

Bernstein, L. R., Ben-Ari, E., Simek, S., and Colburn, N. H. (1991). Gene regulation and genetic susceptibility to neoplastic transformation: AP-1 and p80 expression in JB6 cells. *Environ. Health Perspect.*, 93:111–19.

Bernstein, L. R., and Colburn, N. H. (1989). API/*jun* function is differentially induced in promotion sensitive and resistant JB6 cells. *Science*, 244:566–69.

Beutler, B., and Cerami, A. (1989). The biology of cachection/TNF—A primary mediator of the host response. *Ann. Rev. Immunol.*, 7:625–55.

Boveris, A. (1977). Mitochondrial production of superoxide radical and hydrogen peroxide. In M. Reivich, R. Coburn, S. Lahiri, B. Chance, eds. *Tissue Hypoxia and Ischemia*. New York: Plenum, pp. 67–82.

Brenner, D. A., O'Hara, M., Angel, P., Chojkier, M., and Karin, M. (1989). Prolonged activation of *jun* and collagenase genes by tumour necrosis factor-α. *Nature*, 337:661–63.

Buscher, M., Rahmsdorf, H. J., Litfin, M., Karin, M., and Herrlich, P. (1988). Activation of the c-fos gene by UV and phorbol ester: different signal transduction pathways converge to the same enhancer element. *Oncogene*, 3:1301–11.

Cantoni, O. Sestili, P., Cattabeni, F., Bellomo, G., Pou, S., Cohen, M. et al. (1989). Calcium chelator Quin 2 prevents hydrogen-peroxide-induced DNA breakage and cytotoxicity. *Eur. J. Biochem.*, 182:209–12.

Cerutti, P. A. (1989). Response modification in carcinogenesis. *Environ Health Perspect.*, 81:39–43.

Colburn, N. H., Wendel, E. J., and Abruzzo, G. (1981). Dissociation of mitogenesis and late-stage promotion of tumor cell phenotype of phorbol esters: Mitogen-resistant variants are sensitive to promotion. *Proc. Natl. Acad. Sci. USA*, 78:6912–16.

Crawford, D. R., Amstad, P. A., Yin Foo, D. D., and Cerutti, P. A. (1989). Constitutive and phorbol-myristate-acetate regulated antioxidant defense of mouse epidermal cells JB6. *Mol. Carcinog.*, 2:136–43.

Crawford, D., Zbinden, I., Amstad, P., and Cerutti, P. (1988). Oxidant stress induces the proto-oncogenes c-*fos* and c-*myx* in mouse epidermal cells. *Oncogene*, 3:27–31.

Davidson, N. E., Egner, P. A., and Kensler, T. W., II (1990). Transcriptional control of glutathione S-transferase gene expression by the chemoprotective agent 5-(2-pyrazinyl)-4-methyl-1,2-dithiol-3-thione (oltipraz) in rat liver. *Cancer Res.*, 50:2251–55.

Dixit, V. M., Marks, R. M., Sarma, V., and Prochownik, E. V. (1989). The antimitogenic action of tumor necrosis factor is associated with increased AP-1/c-jun proto-oncogene transcription. *J. Biol. Chem.*, 264:16905–09.

Duke, R. C., Chervenak, R., and Cohen, J. J. (1983). Endogenous endonuclease-induced DNA fragmentation: an early event in cell-mediated cytolysis. *Proc. Natl. Acad. Sci. USA*, 80:6361–65.

Epe, B., Schiffman, D., and Metzler, M. (1986). Possible role of oxygen radicals in cell transformation by diethylstilbestrol and related compounds. *Carcinogenesis*, 7:1329–34.

Fornace, A. J., Alamo, I., and Hollander C. (1988). DNA damage-inducible transcripts in mammalian cells. *Proc. Natl. Acad. Sci. USA*, 85:8800–04.

Fornace, A. J., Nebert, D. W., Hollander, M. C., Luethy, J. D., Papathanasiou, M., Fargnoli, J. et al. (1989). Mammalian genes coordinately regulated by growth arrest signals and DNA-damaging agents. *Mol. Cell. Biol.*, 9:4196–4203.

Fuqua, S. A. W., Blum-Salingaros, M., and McGuire, W. L. (1989). Induction of the estrogen-regulated "24-K" protein by heat shock. *Cancer Res.*, 49:4126–29.

Garrity, R. R., Smith, B. M., and Colburn, N. H. (1989). Genes and signals involved in tumor promoter induced transformation. In N. H. Colburn, ed. *Genes and Signal Transduction in Multistage Carcinogenesis*. New York: Marcel Dekker, pp. 139–66.

Gluecksohn-Waelsch, S. (1987). Regulatory genes in development. *Trends Genet.*, 3:123–27.

Gopalakrishna, R., and Anderson, W. B. (1989). Ca^{2+}- and phospholipid-independent activation of protein kinase C by selective oxidative modification of the regulatory domain. *Proc. Natl. Acad. Sci. USA*, 86:6758–62.

Housey, G. M., Johnson, M. D., Hsiao, W. L. W., O'Brien, C. A., Murphy, J. P., Kirschmeier, P. et al. (1988). Overproduction of protein kinase C causes disordered growth control in rat fibroblasts. *Cell*, 52:343–54.

Kageyama, H., Hiwasa, T., Tokunaga, K., and Sakiyama, S. (1988). Isolation and characterization of a complementary DNA clone for a M_r32,000 protein which is induced with tumor promoters in BALB/c 3T3 cells. *Cancer Res.*, 48:4705–98.

Kensler, T. W., Egner, P. A., Dolan, P. M., Groopman, J. D., and Roebuck, B. D. (1987). Mechanisms of protection against aflatoxin tumorigenicity in rats fed 5-(2-pyranzinyl)-4-methyl-1,2-dithiole-3-thione (oltipraz) and related 1,2-dithiol-3-thiones and 1,2-dithol-3-ones. *Cancer Res.*, 47:4271–77.

Keyse, S. M., and Tyrrell, R. M. (1989). Heme oxygenase is the major 32-kDa stress protein induced in human skin fibroblasts by UVA radiation, hydrogen peroxide, and sodium arsenite. *Proc. Natl. Acad. Sci. USA*, 86:99–103.

Klausner, R. D., and Harford, J. B. (1989). Cis-trans models for post-transcriptional gene regulation. *Science*, 246:870–72.

Kolch, W., Cleveland, J. L., and Rapp, U. R. (1991). Role of oncogenes in the abrogation of growth factor requirements of hemopoietic cells. In W. Paukovits, ed. *Growth Regulation and Carcinogenesis*. Boca Raton FL: CRC Press, pp. 279–304.

Krupitza, G., and Cerutti, P. (1989). ADP-ribosylation of ADPR-transferase and topoisomerase I in intact mouse epidermal cells JB6. *Biochemistry*, 2:2034–40.

Larsson, R., and Cerutti, P. (1988). Oxidants induce phosphorylation of ribosomal protein S6. *J. Biol. Chem.*, 263:17452–58.

Larsson, R., and Cerutti, P. (1989). Translocation and enhancement of phosphotransferase activity of protein kinase C following exposure in mouse epidermal cells to oxidants. *Cancer Res.*, 49:5627–32.

Lin, C. S., Goldthwait, D. A., and Samols, D. (1990). Induction of transcription from the long terminal repeat of Moloney murine sarcoma provirus by UV-irradiation, X-irradiation, and phorbol ester. *Proc. Natl. Acad. Sci. USA*, 87:36–40.

Mello Filho, A. C., Hoffmann, M. E., and Meneghinia, R. (1984). Cell killing and DNA damage by hydrogen peroxide are mediated by intracellular iron. *Biochem, J.*, 218:273–75.

Miskin, R., and Ben-Ishai, R. (1981). Induction of plasminogen activator by UV light in normal and xeroderma pigmentosum fibroblasts. *Proc. Natl. Acad. Sci. USA*, 78:6236–40.

Muehlematter, D., Larsson, R., and Cerutti, P. (1988). Active oxygen induced DNA strand breakage and poly ADP-ribosylation in promotable and non-promotable JB6 mouse epidermal cells. *Carcinogenesis*, 9:239–45.

Nabel, G., and Baltimore, D. (1987). An inducible transcription factor activates expression of human immunodeficiency virus in T cells. *Nature*, 326:711–13.

Nakamura, Y., Colburn, N. H., and Gindhart, T. D. (1985). Role of reactive oxygen in tumor promotion: implication of superoxide anion in promotion of neoplastic transformation in JB6-cells by TPA. *Carcinogenesis*, 6:229–35.

Nakamura, T., Gindhart, T. D., Winterstein, D., Tomita, I., Seed, J. L., and Colburn, N. H. (1988). Early superoxide dismutase-sensitive event promotes neoplastic transformation in mouse epidermal JB6 cells. *Carcinogenesis*, 9:203–07.

Oberley, L. W., Kasemset St. Clair, D., Autor, A. P., and Oberley, T. D. (1987). Increase in manganese superoxide dismutase activity in the mouse heart after X-irradiation. *Arch. Biochem. Biophys.*, 254:69–80.

Oberley, L. W., and Oberley, T. D. (1988). Role of antioxidant enzymes in cell immortalization and transformation. *Mol. Cell. Biochem.*, 84:147–53.

Papathanasiou, M. A., and Fornace, A. J., Jr. (1991). DNA-damage inducible genes. In R. F. Ozols, ed. Molecular and Clinical Advances in Anticancer Drug Resistance. Norwell, MA: Kluwer, pp. 13–36.

Roy, D., and Liehr, J. G. (1989). Changes in activities of free radical detoxifying enzymes in kidneys of male Syrian hamsters treated with estradiol. *Cancer Res.*, 49:1475–80.

Schorpp, M., Mallick, U., Rahmsdorf, H. J., and Herrlich, P. (1984). UV-induced extracellular factor from human fibroblasts communicates the UV response to nonirradiated cells. *Cell*, 37:861–68.

Schütte, J., Minna, J. D., and Birrer, M. J. (1989). Deregulated expression of human c-jun transforms primary rat embryo cells in cooperation with an activated c-Ha-ras gene and transforms Rat-1a cells as a single gene. *Proc. Natl. Acad. Sci. USA*, 86:2257–61.

Seidegard, J., Pero, R. W., Miller, D. G., and Beattie, E. J. (1986). A glutathione transferase in human leukocytes as a marker for the susceptibility to lung cancer. *Carcinogenesis* (London), 7:751–53.

Seidegard, J., Vorachek, W. R., Pero, R. W., and Pearson, W. R. (1988). Hereditary differencs in the expression of the human glutathione transferase active on *trans*-stilbene oxide are due to a gene deletion. *Proc. Natl. Acad. Sci. USA*, 85:7293–97.

Sheehan, D. M., and Branham, W. S. (1987). Dissociation of estrogen-induced uterine growth and ornithine decarboxylase activity in the postnatal rat. *Teratogenesis Carcinog. Mutagen.*, 7:411–22.

Smith, M. R., Munger, W. E., Kung, H-F., Takacs, L., and Durum, S. K. (1990). Direct evidence for an intracellular role for tumor necrosis factor-α. *J. Immunol.*, 144:162–69.

Solis-Herruzo, J. A., Brenner, D. A., and Chojkier, M. J. (1988). Tumor necrosis factor α inhibits collagen gene transcription and collagen synthesis in cultured human fibroblasts. *J. Biol. Chem.*, 263:5841–45.

Stein, B., Kramer, M., Rahmsdorf, H. J., Ponta, H., and Herrlich, P. (1989 Nov). UV induced transcription from the HIV-1 LTR and UV induced secretion of an extracellular factor that induced HIV-1 transcription in non-irradiated cells. *J. Virology*, 63:4540–44.

Stein, B., Rahmsdorf, H. J., Steffen, A., Litfin, M., and Herrlich, P. (1989). UV-induced DNA damage is an intermediate step in UV-induced expression of human immunodeficiency virus type 1, collagenase, c-fos, and metallothionein. *Mol. Cell. Biol.*, 9:5169–81.

Stocker, R., Yamamoto, Y., McDonagh, A. F., Glazer, A. N., and Ames, B. N. (1987). Bilirubin is an antioxidant of possible physiological importance. *Science*, 235:1043–46.

Summers, R. W., Maves, B. V., Reeves, R. D., Arjes, L. J., and Oberley, L. W. (1989). Irradiation increases superoxide dismutase in rat intestinal smooth muscle. *Free Radic. Biol. Med.*, 6:261–70.

Treisman, R. (1985). Transient accumulation of c-fos RNA following serum stimulation requires a conserved 5′ element and c-fos 3′ sequences. *Cell*, 42:889–902.

Valerie, K., Delers, A, Bruck, C., Thiriart, C., Rosenberg, H., Debouck, C. et al. (1988). Activation of human immunodeficiency virus type 1 by DNA damage in human cells. *Nature*, 333:78–81.

Verma, A. K., Hsienh, J. T., and Pong, R. C. (1988). Mechanisms involved in ornithine decarboxylase induction by 12-0-tetradecanoylphorbol-13-acetate, a potent mouse skin tumor promoter and an activator of protein kinase C. *Adv. Exp. Med. Biol.*, 250:273–90.

Wattenberg, L. W., and Bueding, E. (1986). Inhibitory effects of 5-(pyrazinyl)-4-methyl-1,2-dithiol-3-thione(oltipraz) on carcinogenesis induced by benzo(a)-pyrene, diethylnitrosamine, and uracil mustard. *Carcinogenesis* (London), 7:1379–81.

Whitman, S. E., Murch, G., Angel, P., Rahmsdorf, H. J., Smith, B. J., Lyons, A. et al. (1986). Comparison of human stromelysin and collagenase by cloning and sequence analysis. *Biochem. J.*, 240:913–16.

Wong, G. H. W., Elwell, J. H., Oberley, L. W., and Goeddel, D. V. (1989). Manganous superoxide dismutase is essential for cellular resistance to cytotoxicity of tumor necrosis factor. *Cell*, 58:923–32.

7

Iron and Oxidative Damage in Human Cancer

RICHARD G. STEVENS AND KAZUO NERIISHI

Iron is the most abundant transition metal in the human body, and it plays a central role in metabolism. The deleterious effects of severe iron deficiency have been well documented (Addy 1986). However, the potential dangers of iron excess have been examined only insofar as severe iron overload occurs. Increased available body iron stores may increase the risk of cancer by one or both of two possible mechanisms (reviewed in Stevens and Kalkwarf 1990). First, iron can catalyze the production of oxygen radicals, and these may be proximate carcinogens. Iron bound to DNA may be particularly effective in this regard. Second, iron may be a limiting nutrient to the growth and replication of a transformed cell in the human body, and thus, high iron stores may increase the chances that a transformed cell will survive to become a clinically apparent neoplasm. Given the high available iron content of the Western diet, and the fact that dietary practices have been changing throughout the world toward the Western model, the biologic consequences of moderately elevated iron stores deserve attention. Possible consequences are increased risk for cancer and increased sensitivity to radiation injury. Although nutritional antioxidants have received much attention in this regard, the "oxidant" iron has received little.

OXYGEN RADICALS AND IRON

Oxygen radicals include the hydroxyl radical, the perhydroxyl radical, and their deprotonated forms. They are highly toxic species produced intracellularly by reactions that can be catalyzed by iron. These radicals are believed to be some of the primary intermediates in the development of radiobiologic damage. They have also been implicated in the cellular activation of carcinogens, the toxicity of several xenobiotics, and the deleterious effects of aging. Oxygen radicals can damage DNA extensively by inducing strand breakage and degradation of deoxyribose, can induce lipid peroxidation, and may play an important role in carcinogenesis and other disease processes. Halliwell and Gutteridge (1986) discuss the possible role of oxygen radicals in a variety of disease processes and stress the role of iron.

In tissue at pH 7, the hydroxyl radical is primarily protonated, whereas the perhydroxyl radical exists predominantly in its deprotonated form and is called the superoxide radical. This latter radical is formed in almost all aerobic cells. It is removed in vivo by dismutation to hydrogen peroxide and molecular oxygen; this reaction is catalyzed by the copper-zinc enzyme, superoxide dismutase (SOD). Catalase and peroxidases reduce the concentration of the diffusing hydrogen peroxide in the cell by catalyzing its disproportionation.

$$H_2O_2 + H_2O_2 \xrightarrow{\text{Catalase}} 2H_2O + O_2$$

$$2H_2O_2 + 2RH_2 \xrightarrow[\text{peroxidase}]{\text{Glutathione}} 4H_2O + R_2$$

Hydrogen peroxide can react with ferrous (Fe(II)) iron to yield hydroxyl radicals via the Fenton reaction:

$$\text{Fe(II)} + H_2O_2 \longrightarrow \text{Fe(III)} + OH^- + \dot{O}H$$

Transferrin carries ferric iron (Fe(III)) into the cell where it can react with the superoxide radical to form ferrous iron plus molecular oxygen. The net reaction is thus production of hydroxyl radicals inside the cell.

The amount of damage caused by superoxide and hydroxyl radicals depends on their concentration and their chances of diffusing to critical tissue sites. Both radicals react with themselves to form hydrogen peroxide; but, whereas the hydroxyl radical also reacts rapidly with most biochemical structures, the superoxide radical is more selective. The recombination product, hydrogen peroxide, is generally even less reactive and can thus diffuse farther than the superoxide or hydroxyl radicals. However, if ferrous iron is present, hydrogen peroxide can be decomposed to re-form hydroxyl radicals by the Fenton reaction. Thus, the net effect of intracellular ferrous iron is to enhance the likelihood that a reactive hydroxyl radical will reach a critical tissue site.

NONSPECIFIC IRON COMPLEXES

Although neither free ferrous nor ferric ions are thought to exist in tissue in significant concentrations, iron can exist in a variety of iron complexes, for example, Fe(III)ADP (Floyd 1983; Graf et al. 1984). These complexes react rapidly with the superoxide radical to form ferrous iron. This ferrous iron is then available for the Fenton reaction. Graf and colleagues (1984) analyzed the ability of 12 different iron complexes to catalyze the Fenton reaction. They found that, for the iron complex to successfully produce hydroxyl radicals from hydrogen peroxide, there is a requirement for at least one free iron coordination site. For example, although iron complexes of adenosine di- and triphosphate were efficient, iron bound to phytate (naturally occurring in plant fiber) or Desferal (desferrioxamine B methanesulfonate) was ineffective.

Balla and coworkers (1990) loaded cultured endothelial cells with chelated iron using 8-hydroxyquinoline. These iron-loaded cells were extremely sensitive to stimulated granulocytes, menadione, and hydrogen peroxide. As little as 7 μmol/L H_2O_2 resulted in 50% killing of the iron-loaded cells, whereas untreated cells readily survived H_2O_2 concentrations more than 100-fold greater, at 2 mM. The authors concluded that iron sensitized the cells to killing by hydrogen peroxide.

FERRITIN

Excess intracellular iron is stored in ferritin (Jacobs 1985). The ferritin protein consists of 24 subunits for a total molecular weight of approximately 440 000. Ferritin can store up to 4500 atoms of iron as a ferric-oxide-phosphate micelle. The iron stored in ferritin is not necessarily safely sequestered.

Lipid peroxidation may result from a cascade of radical reactions catalyzed by iron complexes in cells (Aust et al. 1985; Samokyszyn et al. 1988). Normally, there is a low concentration of iron capable of catalyzing these events. Under certain circumstances, however, iron associated with binding proteins such as transferrin and ferritin can initiate damage by radical reactions. Wu and colleagues (1990) found evidence that iron overload in rats increased lipid peroxidation. The rats' diet was supplemented with carbonyl iron for 11 to 12 wk. The iron-supplemented diet resulted in increased urinary excretion of thiobarbituric acid-reactive substances and in higher levels of conjugated dienes in hepatic microsomes.

Damage to DNA may also be mediated by ferritin (19% iron by weight). Whiting and colleagues (1981) showed that addition of ferritin to the growth medium of confluent Chinese hamster ovary cells induced chromosome breakage in direct proportion to concentration. Apoferritin (iron-free) had no such effect. The authors concluded that iron-catalyzed radical reactions accounted for the chromosome damage.

A sublethal concentration of ferritin can be a potent radiosensitizer (Nelson and Stevens, unpublished data). Stationary-phase Chinese hamster ovary cells were cultured in medium containing ferritin (approximately

19% iron by weight) added at concentrations ranging from 0 to 128 μg/ml. One set of cultures was unirradiated, and another set was exposed to 4.0 Gy of x-irradiation. Clonogenic cell survival was assessed in each set of cultures. In the absence of added ferritin, 4.0 Gy killed approximately 50% of the cells. In the absence of radiation, ferritin was not toxic at less than 48 μg/ml; above 48 μg/ml, toxicity increased with concentration. Apoferritin was not toxic at any concentration tested (up to 1000 μg/ml). Although 32 μg/ml ferritin was not toxic, it reduced the survival of x-irradiated cells by an additional 75%.

IRON-DNA COMPLEXES

Richter and colleagues (1988) provide evidence that normal oxidative damage to DNA is extensive. They assayed rat hepatocytes for 8-hydroxy-deoxyguanosine (8-OHdG), a measure of oxidative damage to DNA and found a level of 1 per 130 000 bases in nuclear DNA and 1 per 8000 bases in mitochondrial DNA. Since 8-OHdG probably represents a small proportion of oxidative DNA damage, the total damage is more extensive than previously believed. They also found that incubation of hepatocyte mitochondrial DNA with 250 μM Fe^{3+} increased the yield of 8-OHdG threefold.

Imlay and Linn (1988) found that a major source of toxicity of hydrogen peroxide to *Escherichia coli* is mediated by DNA damage from Fenton reactions catalyzed by iron (see Chapter 5). Loeb and coworkers (1988) incubated phage with iron, transfected to the phage *E. coli*, and observed mutation and inactivation of the phage. They speculated that the characteristic damage involves an alteration of deoxyadenosine. Further, the authors extrapolated these results to suggest that free iron localized in cellular DNA may be an important source of mutation to DNA in humans. These reactions are efficient enough to suggest a new technology for DNA cleavage. Oligodeoxynucleotide-EDTA · Fe^{2+} has been used to complex with DNA and cleave at sequence-specific sites nonenzymatically by exploiting Fenton chemistry (Dreyer and Dervan 1985; Chu and Orgel 1985).

Ward and colleagues (1985) provided evidence that double-strand breaks in DNA are required to cause cell death. They use hydrogen peroxide at 0°C to produce single-strand breaks and asserted that these breaks resulted from a metal-catalyzed oxidation of the hydrogen peroxide to the hydroxyl radical. At the cold temperature, the reduction of the ferric back to the ferrous form was slow; thus only one molecule at a time could cause damage. At 37°C hydrogen peroxide was much more lethal. The authors suggested a mechanism whereby hydrogen peroxide diffuses to DNA where it can be oxidized by a DNA-ferrous iron adduct to a hydroxyl radical (Floyd 1981; Lesko et al. 1982; Larramendy et al. 1987); the ferric iron can be reduced once again to ferrous by another hydrogen peroxide molecule, and then react with a third to produce a double-strand break and cell death.

EPIDEMIOLOGIC ISSUES

Studying the possible consequences of these hypothesized biologic mechanisms in humans is difficult. This is particularly difficult in long-term studies of cancer. The first problem is definition of a measurable attribute that assesses iron status and that yields a quantitative indication ranging from "high" to "low," corresponding either directly or inversely to "high" and "low" iron. Quantification of such a measure must be inexpensive in time and money, reproducible, and well tolerated by study subjects. The studies done to date have used serum measures such as ferritin, transferrin, transferrin saturation, and total iron binding capacity (TIBC) before diagnosis of cancer. To interpret the results, assumptions have been made that these measures are related to iron status and that some other underlying condition, such as inflammation, is not invalidating the results. In general, there is a direct correlation of serum ferritin and transferrin saturation with available body iron stores and an inverse correlation of serum transferrin and TIBC with iron stores; that is, high iron stores result in high ferritin and transferrin saturation and in low serum transferrin and TIBC.

Although there is a wide range of body iron levels among healthy individuals, as reflected in serum ferritin levels, it is not clear the extent to which dietary iron intake influences these levels within the normal range (Cook 1990). Iron in diet is in two forms: heme iron and inorganic, or nonheme, iron. The mechanisms for regulating absorption of nonheme iron are more efficient than the mechanisms for regulating heme iron uptake. The increase in heme iron consumption in industrialized societies, therefore, may lead to body iron stores above those required for optimal health.

STUDIES DONE TO DATE

The results of four published studies are reviewed by Stevens (1990) and are compared in Figure 7-1, along with results from a new study of stomach cancer in Japanese atomic bomb survivors (Akiba et al. 1991). The methods of the reported studies are similar. Each is based on a study population followed prospectively.

Stevens and coworkers (1983) studied the relationship of serum ferritin and transferrin to subsequent risk of death over a 10-year period in the Solomon Islands. Outcome in this study was all-cause mortality, which probably did not include a large proportion of cancer deaths. Six subgroups

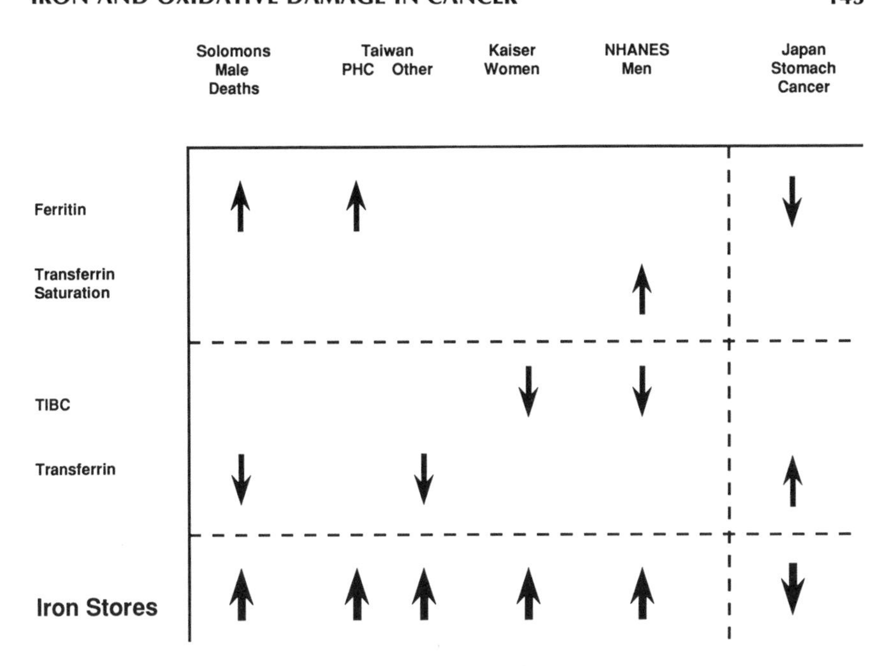

Figure 7–1 Summary of five studies of iron stores and cancer risk. For men in the Solomon Islands, Taiwan, and NHANES studies, and for women in the Kaiser study, results were consistent with the hypothesis that high body iron stores increase overall cancer risk. For stomach cancer patients in the Japan study, results were in the opposite direction. The direction of the arrow indicates whether the "cases" had higher or lower levels.

of Solomon Islanders were first examined during the period 1966 to 1970 as part of the Harvard–Solomon Islands project (Friedlaender 1987). The overall objective of the project was to define a longitudinal sample of the indigenous Solomon Islands population and to investigate the subsequent effects of Westernization. At the start of the study a blood sample was taken; demographic characteristics were recorded; a medical examination was performed; and anthropometric measurements were taken on each of approximately 2500 individuals. The alive-dead status of each of these people was determined as of 1978 by anthropologists in the Solomon Islands. The stored serum of each person who had died and that of a selected age–sex-matched control person who had not died, were tested for ferritin and transferrin. There was no adjustment for smoking.

From 1975 to 1978, 21 513 male Chinese government workers in Taiwan were enrolled in a study (Beasley et al. 1981) of the effect of hepatitis B virus on risk for primary hepatocellular carcinoma (PHC). Stevens and coworkers (1986) reported the results from a study of serum ferritin and transferrin levels in serum, stored since enrollment, from 192 of these men who developed PHC or who died of any cancer by 1983, and 358 age-matched control men who had not died or developed cancer. There was no adjustment for smoking.

Selby and Friedman (1988) reported on over 175 000 members of the

Kaiser health plan in northern California, who were followed from 1964 to 1973. Each subject had a baseline medical examination during this period and a TIBC measurement. The group was followed through 1980, and incident cases of cancer were recorded in the large database. TIBC was compared between cases and those who did not develop cancer over the study period. Subjects who developed cancer within 2 years of the blood tests were excluded from the analyses, and adjustment was made for age and smoking.

Utilizing the existing database of the National Health and Nutrition Examination Survey I (NHANES I) in the United States, Stevens and coworkers (1988) compared transferrin saturation and TIBC in those of the 14 707 subjects who developed cancer as of 1984 and those who did not. The cohort was identified during the period 1971 to 1975 as a probability sample of the United States adult population. An extensive dietary questionnaire was administered; a medical examination was performed; anthropometric measurements were made; and blood was taken and analyzed for a large number of constituents. Subjects who developed cancer within 4 years of the time the blood was taken were not included in the analysis, and adjustment was made for age and smoking. The blood was not saved.

A recent study (Akiba et al. 1991) was completed in which serum ferritin and transferrin levels were correlated with the risk of stomach cancer in Japanese atomic bomb survivors. From 1970 to 1972, blood was taken from members of the Adult Health Study of atomic bomb survivors at their biennial clinical examination at the Radiation Effects Research Foundation in Hiroshima and Nagasaki. There were 233 cases of stomach cancer diagnosed in this group from 1973 to 1983. Serum ferritin and transferrin levels in the serum stored from 1970 to 1972 for the patients in the case group were compared to the levels from subjects in an age-, sex-, city-, and radiation-matched control group. Adjustment was made for smoking.

The groups used for each of these studies were different; the metrics for assessment of iron status differed somewhat; and the outcomes were different. However, comparison of results may provide further insight into the possible effect of iron status on cancer risk. Serum ferritin and serum transferrrin levels were used in the Solomon Island study of general mortality, in the Taiwan study of incidence of PHC and cancer death excluding PHC, and in the Japanese study of stomach cancer incidence. The NHANES study used transferrin saturation and TIBC, whereas the Kaiser study used only TIBC.

The direction of each arrow in Figure 7-1 indicates whether the mean among the cases (cancer or death over the study period) was higher or lower than the mean among the control subjects. The arrow associated with "iron stores" is in the direction inferred from the results of the actual measurements of ferritin, transferrin, transferrin saturation, or TIBC. Again, in general, high iron stores result in high ferritin and transferrin saturation and in low serum transferrin and TIBC.

The Solomon Island (Stevens et al. 1983), Taiwan (Stevens et al. 1986), and NHANES (Stevens et al. 1988) studies have generated results consistent with the hypothesis that higher iron stores increase risk of death or cancer in men. Findings from the Kaiser study (Selby and Friedman 1988) support an association in women but not in men. It was suggested that women in the NHANES study with high transferrin saturation levels might also be at moderately elevated risk of cancer.

In the Japan study (Akiba et al. 1991), the association of serum ferritin and transferrin is in the opposite direction: lower iron is associated with an increased incidence of stomach cancer. There were only 8 cases of stomach cancer in the NHANES study and 24 in the Taiwan study. However, the 8 men who developed stomach cancer in the NHANES study had a transferrin saturation level of 26.4, lower than the control subjects, and TIBC of 67.0, which was higher than that of the control subjects (not shown in Figure 7-1). In the Taiwan study, subjects with stomach cancer also had lower ferritin levels than did control subjects. Thus, the results of these three studies are consistent with each other when restricted to analysis of stomach cancer. A mechanism whereby low stomach acidity leads to both low iron stores and stomach cancer is proposed in Akiba and coworkers (1991) to account for the stomach cancer results in these studies.

Dietary iron intake was examined in the NHANES study and found not to be related to cancer risk. However, the estimate of iron intake may not have been accurate because it was based on one 24-h dietary recall questionnaire developed in the early 1970s. Dietary assessment instruments have improved since that time. In a recent paper, Freudenhein and coworkers (1990) reported that high dietary iron intake was associated with increased risk of rectal cancer in a case-control study from New York State. This result is consistent with the hypothesis of Graf and Eaton (1985), who state that dietary phytate may reduce risk of colon cancer by virtue of its binding of iron such that it is unavailable for Fenton reactions.

Lowering available iron might be a useful treatment for cancer patients based on the Weinberg (1984) hypothesis that cancer cells depend on iron to flourish. This possibility is reviewed by Cazzola and colleagues (1990).

NATURALLY OCCURRING ANTIOXIDANTS AND CANCER RISK

The relationship of antioxidants such as selenium, carotene, and vitamins E and C to risk for cancer was reviewed in recent books by the National Research Council (1989), the Surgeon General (1988), and Moon and Micozzi (1989), and is discussed by Kensler and Guyton (see Chapter 8). Evidence consists of experimental studies in animals and two kinds of studies in humans: epidemiologic and interventional. The epidemiologic studies of micronutrients and cancer have relied either on reported intake of foods or on tests, using stored serum samples, of the level of the nutrient in question. Both experimental and epidemiologic investigations of specific antioxidants and cancer are suggestive but not conclusive. Whereas some epidemiologic studies support a protective effect of vitamins C and E and selenium, others show no relationship between either the serum level or the estimated dietary intake of each of these micronutrients and risk for cancer.

Retinol and β-carotene have been proposed as cancer-protective nutrients, and a number of epidemiologic studies have been performed to test this hypothesis (Micozzi 1989). Perhaps the strongest evidence of this relationship would be provided by comparing the serum levels of these nutrients in patients with cancer and in control subjects. Better still would be to test serum samples that had been collected long before the cancer had appeared. Ten prospective studies of well-nourished groups have been reported in the literature. No effect was seen in seven of these for retinol; three of five reported lower β-carotene in cases of lung and stomach cancers than in controls; and one study of total serum carotene found no effect. Taken together these studies do not support the hypothesis that high blood levels of retinols reduce the risk for cancer. Epidemiologic findings are inconsistent with β-carotene as well.

The rationale for suspecting that vitamin A and provitamin A (β-carotene) might reduce risk for cancer comes from two sources. First, experimental studies that began in 1925 have shown a link between vitamin A and the appearance of neoplasia (Lippman and Meyskens 1989). Vitamin A is required for the growth and differentiation of normal cells, and, in some systems, has been shown to cause redifferentiation of neoplastic cells. Second, over 25 epidemiologic studies have shown a strong and consistent inverse association between the dietary consumption of green, leafy vegetables (for example, broccoli, brussels sprouts, cabbage, kale, and spinach) and risk for cancer. The assumption has been that the β-carotene content of these vegetables accounts, in part, for their cancer protection capability. However, Micozzi and colleages (1990) analyzed these vegetables and found that only 10% to 20% of the total carotenoid content was β-carotene; 80% to 90% was oxygenated carotenes (for example, lutein), which do not have vitamin A activity. Vegetables that are high in β-carotene (for example, carrots and sweet potatoes) have not been consistently associated with reduced risk for cancer. Also of significance is the fact that consumption of animal products high in vitamin A has not been shown to reduce cancer risk. Micozzi and coworkers (1990) speculate that the cancer-preventive capability of carotenes (including both carotenes with vitamin A activity and those without) results from their ability to scavenge oxygen radicals and not from their vitamin A activity.

Because researchers recognized the inherent limitations of previous studies, they have undertaken intervention trials. For example, an intervention trial is underway in Washington State to test the hypothesis that oral β-carotene or retinol or both will reduce the risk for lung cancer in smokers and asbestos-exposed individuals (Omenn 1988). Several thousand subjects are being recruited and randomly assigned to receive capsules of β-carotene or retinol or both or placebo. The study is a double-blind randomized trial in which neither the subject nor the researchers know who is receiving vitamin or placebo. It is believed that only such a trial will convincingly demonstrate an effect of a micronutrient on the risk for cancer or document the lack of effect.

An important problem with all these studies is the recognition that the effect of any single nutrient may depend heavily on the levels of other nutrients. Thus, studies of antioxidants should include assessment of levels of "oxidants" such as iron. There may be important interactions of the effects of each with the others.

SUSCEPTIBILITY TO OXIDATIVE DAMAGE

Biologic Mechanisms of Defense

An intriguing finding from the studies of the relation between iron and cancer (Stevens et al. 1983, 1986, 1988) has been the statistically significant inverse association of serum albumin and risk for death: Study subjects who have died or developed cancer have had lower albumin levels long before death or diagnosis than those who have survived and remained free of cancer. Serum albumin has long been used to assess protein-calorie nutritional status (Haider and Haider 1984). Its half-life of about 20 days makes it insensitive to immediate changes in protein intake, but a good indicator of cumulative intake in the recent past. Serum albumin is important for osmotic regulation of blood and in many transport and regulatory processes (Peters 1985). It binds an impressive array of substances including fatty acids, heme, calcium, and transition metals such as nickel, copper, and zinc. It also binds tryptophan, thyroxine, and many therapeutic drugs. Despite the fact that serum albumin has been extensively studied for many years, its three-dimensional structure has only recently been eluciated (Carter et al. 1989).

Phillips and coworkers (1989) reported that death rates from *both* cardiovascular disease and cancer were significantly higher in men with low albumin levels (measured at least 5 years prior to death) and that there was also a dose-response relation: the lower the albumin the higher the death rate. The authors stated that although they did not have an *a priori* hypothesis about serum albumin, the strength of the association rivals that between cigarette smoking and mortality and is therefore important to investigate. Hasuo and colleagues made a similar observation in 1986.

Stevens and Blumberg (1990) offered several possible biologic mechanisms that might account for the association of low levels of albumin and increased risk for death from cancer and cardiovascular disease. In Stevens and coworkers (1986, 1988), the effect of albumin on siderophore-mediated iron acquisition was cited as one possible explanation for an association with cancer (Konopka and Neilands 1984). If albumin affects iron availability, then low levels of albumin may result in more available iron for cancer cells to consume or for Fenton chemistry reactions in tissue that might create cancer cells. Oxidative mechanisms may be a common cause of both cancer and cardiovascular disease (Cross 1987). With that in mind, albumin levels may affect oxidative stress by another mechanism. Conjugated linoleic acids (CLAs) are derivatives of linoleic acid that contain a conjugated double bond (Pariza 1988). They are potent antioxidants and are effective anticarcinogens in animals. CLAs are found in human serum, bile, and duodenal juice. The source is unknown. Examples of foods that are high in CLAs are grilled beef (Ha et al. 1987) and processed cheese (Ha et al. 1989). The production of CLAs from linoleic acid requires protein, the most abundant of which in serum is albumin (Dormandy and Wickens 1987; Cawood et al. 1983). Therefore, albumin may be quantitatively related to CLA production in the body; low albumin may lead to low CLA, which, in turn, may result in increased oxidative stress and its possible long-term sequelae, cancer and cardiovascular disease. Another possibility is that low serum albumin levels may reflect a low intake of protein-rich foods that are high in CLAs, which may reduce risk of disease, and thereby be indirectly related to cancer and cardiovascular disease mortality.

Apart from the possible effects of albumin on such substances as iron and CLAs, albumin itself may have antioxidant capacity. Halliwell (1988) proposed that the protein may behave as a "sacrificial antioxidant." It is present in such large quantities that small amounts may scavenge radicals, be functionally destroyed, and lower the level of oxidative stress, while not significantly affecting the pool of functional albumin.

The observation of a strong inverse relation between serum albumin and chronic disease mortality is open to a variety of interpretations and should be vigorously investigated.

Genetic Susceptibility: Hemochromatosis

Hemochromatosis is a genetically determined disease in which body iron stores increase to the point of causing serious health consequences such as liver cancer and cardiovascular disease. The hemochromatosis gene is located on chromosome 6, probably between the HLA-A and B loci (Edwards et al. 1986). The HLA A3, B7, and B14 antigens are found much more frequently in hemochromatosis patients than expected (Bregman et al. 1980), and the relative risk of hemochromatosis associated with HLA A3 antigen has been estimated to be as high as 8.2 with an etiologic fraction of 0.67 (Svejgaard et al. 1983). Kravitz and coworkers (1979) reported a transferrin saturation of 93% and 83.7% in hemochromatosis homozygous men and women, respectively, compared with only 32.8% and 30.2% in homozygous normal men and women. Obligate heterozygotes had an intermediate transferrin saturation of 42.6% and 40.1%, respectively. Because the gene frequency of the abnormal hemochromatosis allele has been estimated from a study of blood donors in Utah to be 0.067 (Edwards et al. 1988), approximately 12% to 14% of the population may be heterozygous. This calculation raises the possibility that a large segment of the normal population may be at genetically determined increased risk of oxidative stress because of chronically higher body iron stores (for example, higher transferrin saturation) resulting from the hemochromatosis gene (in either the homozygous or heterozygous state).

Patients with primary hemochromatosis are at greatly increased risk for death from liver cancer, cardiomyopathy, and cirrhosis (Niederau et al. 1985). Iron stores also increase in patients given multiple transfusions to treat severe anemias. For example, patients with thalassemia die at an early age of heart disease, infection, liver disease, or cancer (Zurlo et al. 1989). Treatment with the iron chelator desferrioxamine was initiated in 1975 in Italy, and since that time, survival of thalassemic children has improved significantly. Thus, in both hemochromatosis and thalassemia, iron overload can induce severe physical harm.

Radiation Sensitivity

Individual differences in genetic susceptibility to oxidative stress may be most profitably studied in groups exposed to a severe oxidant stressor such as toxic chemicals or radiation. The ability to scavenge radicals may vary considerably among individuals, and this variability may be genetically determined. In extreme cases, for example, an enzyme such as catalase may be absent. More subtle quantitative differences in antioxidant enzyme levels may also play an important role in protecting tissues from the ravages of chronic exposure to oxygen radicals or to an extreme-point exposure such as an atomic bomb blast.

Survivors of atomic bomb blasts show some signs of subclinical inflammation such as increased neutrophil counts, accelerated erythrocyte sedimentation rate, and increased levels of acute phase proteins (Sawada et al. 1986; Neriishi et al. 1986). Acute phase proteins are known to be induced by cytokines such as interleukin 1 (IL-1), interleukin 6 (IL-6), or tumor necrosis factor (TNF) (Dinarello 1985; Andus et al. 1988; Nishimoto et al. 1989: Castell 1988). Whereas IL-6 and TNF stimulate polymorphonuclear cells (PMN) to generate superoxide (Ozaki et al. 1987; Berkow et al. 1987), superoxide activates monocytes and PMN to produce IL-1 (Kasama et al. 1989; Gougerot-Pocidalo et al. 1989). Cytokine-oxidative metabolite cyclic reaction may lead to oxidative stress to the body, which may be operating in atomic-bomb survivors.

Further, because there is evidence that immunoglobulins are vulnerable to free radicals (Griffiths and Lunec 1989) and that aggregated immunoglobulins stimulate neutrophils to generate superoxide (Lunec 1984), a chain reaction might occur leading to chronic superoxide generation. Because superoxide modifies lymphocyte proliferation (Duncan and Lawrent 1989) and inhibits T suppressor cells (Zoscheke and Staite 1987) and because hydrogen peroxide induces Epstein-Barr (EB) virus production (Oya et al. 1987), lymphocyte-related abnormalities such as decreased PHA response (Yamakido et al. 1983) and increased infection rate with EB virus (Kumagai et al. 1985) and hepatitis B virus (Kato et al. 1983) observed in atomic-bomb survivors are possibly induced by oxidative stress. Taking into account the fact that chronic stimulation with IL-6 or TNF leads to development of multiple myeloma (Kawano et al. 1988; Bataille et al. 1989; Hata et al. 1990) as well as acute phase protein production, the high incidence of multiple myeloma in atomic-bomb survivors (Ichimaru et al. 1982) could be caused by subclinical oxidative stress.

Antioxidant Enzymes

Hydrogen peroxide may play an important role in the differential effect of radiation on somatic cells. The possible role in radiation damage of antioxidant enzymes such as SOD, catalase, and glutathione perioxidase (GP) may be related to direct or indirect effects on hydrogen peroxide.

Addition of CuZn-SOD to the growth medium of mouse epidermal JB6 cells profoundly reduces the promotion of neoplastic transformation by TPA (Nakamura et al. 1988). These authors speculate that the mechanism of TPA-induced promotion depends on the generation of oxygen radicals. SOD has also been reported to protect against radiation damage (Lipecka et al. 1984; Petkau et al. 1976), but other studies have not shown a protective effect. On the contrary, Scott and coworkers (1989) reported that SOD *amplifies* sensitivity of *E. coli* to ionizing radiation. They suggest that altering, or raising, only SOD while holding other antioxidant systems constant leads to accumulation of H_2O_2, which is toxic. The role of SOD among radiosensitive diseases is not clear. Levels of SOD in Bloom syndrome, Down syndrome, and Parkinson's disease are increased whereas levels are decreased in xeroderma pigmentosum (Nicotera et al. 1989; Otsuka et al. 1985; Saggu et al. 1989; Marttila 1988; Nishigori et al. 1989). Hydrogen peroxide, generated by SOD, may preserve radical-damaging capability in a form that can diffuse through membranes to a site such as a DNA-iron complex where it can be oxidized to hydroxyl radicals and effect significant damage (Stevens and Kalkwarf 1990).

Levels of catalase are reported to be decreased among radiosensitive diseases such as xeroderma pigmentosum, ataxia telangiectasia (AT4BI, AT5BI), retinoblastoma, and Wilms tumor (Vuillume et al. 1986; Crawford et al. 1988; Sabatier et al. 1989; Junien et al. 1980). These diseases also reflect a sensitivity to oxidative stress, UV light, or x-irradiation. Red blood cells from individuals with acatalasemia who do not have serious clinical complications are sensitive to radiation (Aebi et al. 1968), and mammalian epithelial cells are sensitized to hydrogen peroxide by radiation (Link 1987). This finding suggests that catalase may have some protective effect against the oxidative effects of radiation. Because catalase and glutathione peroxidase are inhibited by superoxide (Kono and Fridovich 1982; Blum and

Fridovich 1985), the protective effect of catalase may be limited by overproduction of superoxide.

Glutathione metabolism, which is important in detoxifying hydrogen peroxide, has an important role in protecting the cell from oxidative stress. Increased sensitivity to radiation is observed in human fibroblasts with glutathione synthetase deficiency (Debieu et al. 1985) and in glutathione-depleted cells (Bump et al. 1986). Increased sensitivity to chemical oxidative stress is also observed in one isotype of glutathione transferase (Seidegard et al. 1985) and in cells impaired in glutathione redox cycling (Nathan et al. 1980). However, selenite-induced glutathione peroxidase failed to protect human cells from ionizing radiation damage (Sandstrom et al. 1989a, 1989b). Whereas depletion of glutathione radiosensitizes mouse kidney, depletion of glutathione peroxidase does not (Stevens et al. 1989). In adriamycin-resistant cell lines (GLC4-Adr1 and GLC4-Adr2), which are also radiation resistant, the levels of SOD, catalase, and glutathione peroxidase are not increased (Meijer et al. 1987).

Hydrogen peroxidase has an important role in the pathologic mechanism of oxidative stress, and this effect is modified by three antioxidant enzymes. Studies are underway to investigate differential detoxifying systems for hydrogen peroxide in atomic-bomb survivors in relation to radiosensitivity.

A close relation between epilation and leukemia mortality was observed even after adjustment for reported radiation dose. This finding suggests the possibility of either a radiosensitive subpopulation among atomic-bomb survivors or estimation errors of radiation dose (Neriishi et al. 1989). A similar relation was observed between epilation and chromosome aberration in atomic-bomb survivors (Sposto et al. 1990), although experimental studies in lymphocytes failed to show individual differences in radiosensitivity (Ban et al. 1990; Nakamura et al. 1989).

Nakayama and colleagues (1989) reported a quantitative relationship between tar content of different brands of cigarette and yield of H_2O_2. Thus, the body iron level of a smoker may influence the pathogenic effects of smoking. If there are radiosensitive individuals among atomic-bomb survivors, pharmacogenetic studies could be done using response to smoking as an indicator of oxidative stress (Helman and Rubinstein 1975; Stram et al. 1989). Individuals who respond to cigarette smoke strongly (as seen in leukocyte count and levels of acute phase proteins) may be those who suffered epilation or other short-term effects of radiation, since smoking and radiation may share some aspects of biologic damage mechanisms through induction of oxidative stress (Nakayama et al. 1989; Hoidal and Niewoehner 1982; Van Rensburg et al. 1989; Yada et al. 1985). This study is being planned.

CONCLUSION

Evidence that iron plays a role in oxidative stress and cancer induction is growing. In addition, a common mechanism for cancer and cardiovascular disease may be oxidative stress. Sullivan (1989) suggested that the difference in risk of heart disease between men and women results, in part, from differences in iron stores; the lower the iron stores, the lower the risk. He further suggested that aspirin may reduce risk of heart disease by depleting the body of iron (Sullivan 1982). Studies of iron status and risk of heart disease are underway to test Sullivan's hypotheses.

Figure 7-2 depicts a dichotomy between oxidative stress resulting from naturally occurring oxidants and antioxidants and oxidative stress resulting from the action of xenobiotics. Genetics plays an important role in both mechanisms of stress. Single-gene conditions such as hemochromatosis and polygenic determination of ambient concentration of antioxidant defense enzymes will contribute to susceptibility to disease resulting from oxidative stress.

Similar chemical mechanisms of damage to biomolecules by oxygen radicals may operate in all types of cells and tissues (Figure 7-3), but this damage may manifest itself differently depending upon the growth characteristics of the tissues. If the tissue is slowly dividing (endothelium and myocardium), then perhaps accumulated damage to structural elements leads to tissue dysfunction that results in heart disease and death. A DNA lesion in the cells of these tissues may go unnoticed because of lack of cell division and lack of the opportunity for "fixation" to a transformed phenotype.

In a rapidly dividing tissue (epithelium), structural damage may not be so important to tissue function because of cell turnover and the introduction of new healthy cells. DNA lesions that could lead to cancer (perhaps an "initiating event") may have a greater probability of becoming fixed in at least one cell of the tissue because of cell division.

Thus, rapid turnover of cells may protect a tissue from dysfunction but increase the risk of transformation. Slow turnover of cells may protect a tissue from malignant transformation because of low probability of fixation at cell division of a critical DNA lesion, but may increase the risk of accumulated damage leading to dysfunction.

Iron-mediated increases in oxidative stress may be important in the promotion mechanisms discussed by Colburn (see Chapter 6). Ferritin iron may increase chances for changes in DNA leading to activation of oncogenes or inactivation of antioncogenes leading to cancer (for example, by promotion through increased oxidative stress). In addition, lipid peroxidation by ferritin-iron may increase cardiovascular damage and lead to cardiovascular disease.

DNA from tumor tissue may be assayed for iron-mediated damage, thus providing evidence for or against a role for DNA-iron adducts in the pathogenesis of cancer. Electron spin resonance spectroscopy for radicals in biologic material (for example, nails and hair) discussed by Mason (Chapter 2) may prove useful as a summary measure of oxidative stress and may be exploited in epidemiologic studies of the larger question of whether, and to what extent, oxygen radicals and oxidative stress are important in the

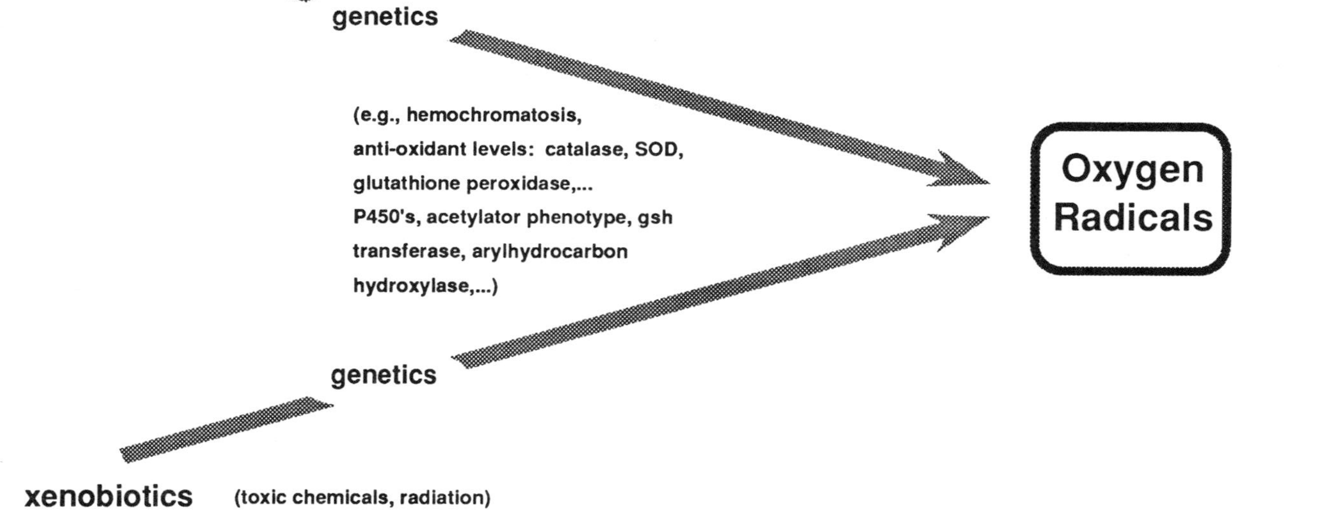

Figure 7–2 Oxygen radicals may be generated by a variety of mechanisms, both as a natural result of respiration and from xenobiotics. The toxicity of a xenobiotic may result primarily from generation of oxygen radicals or only secondarily from their production. Genetics undoubtedly plays a role in all aspects of oxygen radical toxicity.

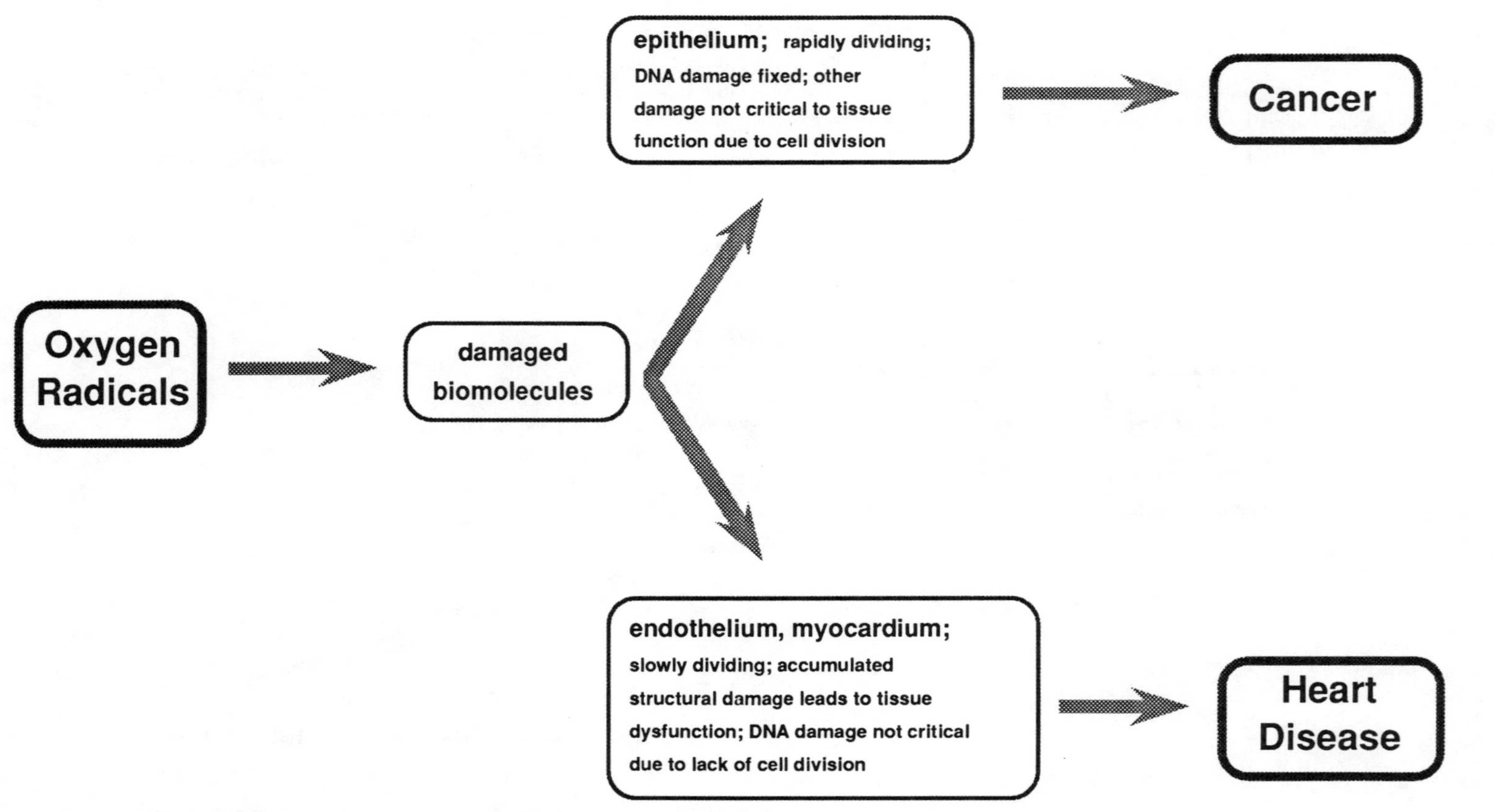

Figure 7–3 Proposed relation between oxygen radicals and the genesis of cancer and heart disease. Biomolecular damage from oxygen radicals may be similar in all types of tissues but manifest itself in very different diseases depending on the rapidity of cell division in each tissue. Slow cell division may protect from consequences of DNA damage but lead to susceptibility to tissue dysfunction; rapid cell division may protect from the consequences of accumulated damage on tissue function, but lead to susceptibility to manifestations of DNA damage such as transformation.

pathogenesis of cancer and cardiovascular disease. The oxy-LDLs discussed by Chisolm (see Chapter 4) and the CLAs studied by Pariza (1988) may interact with each other in the pathogenesis of cardiovascular disease.

Genetic susceptibility to oxidative damage may in part result from genetic variations in the absorbtion (for example, hemochromatosis gene product), transport (for example, transferrin variants), or intracellular processing (for example, ferritin variants) of iron. Whereas antioxidants and their metabolism have received a great deal of attention for their putative ability to protect cells and tissue from oxidative damage, the important oxidant iron has received relatively little. In particular, interactions of effects of iron with antioxidants should be investigated. A cell under oxidative stress by virtue of excess iron may react poorly to challenge by a chemical or radiation at a level easily accommodated by an unstressed cell. Genetic variation in oxidant protection mechanisms may also play an important role in susceptibility to the deleterious effects of excess iron, chemicals, radiation, and aging.

REFERENCES

Addy, D. P. (1986). Happiness is: iron. *Br. Med. J.*, 292:969–70.

Aebi, H., Bossi, E., Cantz, M., Matsubara, S., and Suter, H. (1968). Acatalasemia in Switzerland. In E. Beutler, ed., *Hereditary Disorders of Erythrocyte Metabolism*, New York: Grune & Stratton, pp. 41–65.

Akiba, S., Neriishi, K., Blot, W. J., Kabuto, M., Stevens, R. G., Kato, H. et al. (1991). Serum ferritin and stomach cancer risk among Japanese. RERF TR 14-89 and *Cancer*, 67:1707–12.

Andus, T., Geiger, T., Hirano, T. et al. (1988). Action of recombinant human interleukin 6, interleukin 1β and tumor necrosis factor α on the mRNA induction of acute phase proteins. *Eur. J. Immunol.*, 18:739–46.

Aust, S. D., Morehouse, L. A., and Thomas, C. E. (1985). Role of metals in oxygen radical reactions. *J. Free Radic. Biol. Med.*, 1:3–25.

Balla, G., Vercellotti, G. M., Eaton, J. W., Jacob, H. S. (1990). Iron loading of endothelial cells augments oxidant damage. *J. Lab. Clin. Med.*, 116:546–54.

Ban, S., Setlow, R. B., Bender, M. A. et al. (1990). Radiosensitivity of skin fibroblasts from atomic bomb survivors with and without breast cancer. RERF TR 6-90.

Bataille, R., Jourdan, M., Zhang, X. G., and Klein, B. (1989). Serum levels of interleukin-6, a potent myeloma cell growth factor, as a reflection of disease severity in plasma cell dyscrasias. *J. Clin. Invest.*, 84(6):2008–11.

Beasley, R. P., Hwang, L.-Y., Lin, C.-C. et al. (1981). Hepatocellular carcinoma and hepatitis B virus. *Lancet*, 2:1129–33.

Berkow, R. L., Wang, D., Larrick, J. W., Dodson, E. W., and Howard, T. H. (1987). Enhancement of neutrophile superoxide production by preincubation with recombinant human tumor necrosis factor. *J. Immunol.*, 139:3783–91.

Blum, J., and Fridovich, I. (1985). Inactivation of gluthathione peroxidase by superoxide radical. *Arch. Biochem. Biophys.*, 240:500–08.

Bregman, H., Gelfand, M. C., Winchester, J. F. et al. (1980). Iron-overload-associated myopathy in patients on maintenance haemodialysis: a histocompatibility-linked disorder. *Lancet*, 2:882–85.

Bump, E. A., Jacobs, G. P., Lee, W. W. et al. (1986). Radiosensitization by diamide analogs and arsenicals. *Int. J. Radiat. Oncol. Biol. Phys.*, 12:1533–35.

Carter, D. C., He, X.-M., Munson, S. H., Twigg, P. D., Gernert, K. M., Broom, B. et al. (1989). Three-dimensional structure of human serum albumin. *Science*, 244:1195–98.

Castell, J. U. (1988). Recombinant human interleukin 6 (IL-6/BSF-2/HSF) regulates the synthesis of acute phase proteins in human hepatocytes. *FEBS Lett.*, 232:347–50.

Cawood, P., Wickens, D. G., Iversen, J. M., Braganza, J. M., and Dormandy, T. L. (1983). The nature of diene conjugation in human serum, bile, and duodenal juice. *FEBS Lett.* 162:239–43.

Cazzola, M., Bergamaschi, G., Dezza, L. et al. (1990). Manipulations of cellular iron metabolism for modulating normal and malignant cell proliferation: achievements and prospects. *Blood*, 75:1903–19.

Cook, J. D. (1990). Adaptation in iron metabolism. *Am. J. Clin. Nutr.*, 51:301–08.

Crawford, D., Zbinden, I., Monet, R., Cerrutti, P. (1988). Antioxidant enzymes in xeroderma pigmentosum fibroblasts. *Cancer Res.*, 48:2132–34.

Cross, C. E., (moderator). (1987). Oxygen radicals and human disease. *Ann. Intern. Med.*, 107:526–45.

Debieu, D., Deschavanne, P. J., Midander, J., Larsson, A., and Malaise, E. P. (1985). Survival curves of glutathione synthetase deficient human fibroblasts: correlation between radiosensitivity in hypoxia and glutathione synthetase activity. *Int. J. Radiat. Biol.*, 48:525–43.

Dinarello, C. A. (1985). Interleukin 1 and the pathogenesis of the acute phase response. *N. Engl. J. Med.*, 311:1413–18.

Dormandy, T. L., and Wickens, D. G. (1987). The experimental and clinical pathology of diene conjugation. *Chem. Phys. Lipids*, 45: 353–64.

Dreyer, G. B., and Dervan, P. B. (1985). Sequence-specific cleavage of single-stranded DNA: oligodeoxynucleotide-EDTA·Fe(II). *Proc. Natl. Acad. Sci. USA*, 82:968–72.

Duncan, D. D., Lawrent, D. A. (1989). Differential lymphocyte growth-modifying effects of oxidants: Changes in cytosolic Ca2+. *Toxicol. Appl. Pharmacol.*, 100:485–97.

Edwards, C. Q., Griffen, L. M., Dadone, M. M. et al. (1986). Mapping the locus for hereditary hemochromatosis: localization between HLA-B and HLA-A. *Am. J. Hum. Genet.*, 38:805–11.

Edwards, C. Q., Griffen, L. M., Goldgar, D. et al. (1988). Prevalence of hemochromatosis among 11,065 presumably healthy blood donors. *N. Engl. J. Med.*, 318:1355–62.

Floyd, R. A. (1981). DNA-ferrous iron catalyzed hydroxyl free radical formation from hydrogen peroxide. *Biochem. Biophys. Res. Commun.*, 99:1209–15.

Floyd, R. A. (1983). Direct demonstration that ferrous ion complexes of di- and tri-phosphate nucleotides catalyze hydroxyl free radical formation from hydrogen peroxide. *Arch. Biochem. Biophys.*, 225:263–70.

Freudenheim, J. L., Graham, S., Marshall, J. R. et al. (1990). A case-control study of diet and rectal cancer in western New York. *Am. J. Epidemiol.*, 131:612–24.

Friedlaender, J. S., ed. (1987). *The Solomon Islands Project: A Long-Term Study of Health, Human Biology, and Culture Change*. Oxford: Clarendon Press.

Graf, E., and Eaton, J. W. (1985). Dietary suppression of colonic cancer: fiber or phytate? *Cancer*, 56:717–18.

Graf, E., Mahoney, J. R., Bryand, R. G. et al. (1984). Iron-catalyzed hydroxyl radical formation. *J. Biol. Chem.*, 259:3620–24.

Gougerot-Pocidalo, M. A., Roche, Y., Fay, M., Perianin, A., and Bailly, S. (1989). Oxidative injury amplifies interleukin-1-like activity produced by human monocytes. *Int. J. Immunopharmacol.* 11(8):961–69.

Griffiths, H. R., and Lunec, J. (1989). The effect of free radicals on the carbohydrate moiety of IgG. *FEBS Lett.*, 245:95–99.

Ha, Y. L., Grimm, N. K., and Pariza, M. W. (1987). Anticarcinogens from fried beef: heat-altered derivatives of linoleic acid. *Carcinogenesis*, 8:1881–87.

Ha, Y. L., Grimm, N. K., and Pariza, M. W. (1989). Newly recognized anticarcinogenic fatty acids: identification and quantification in natural and processed cheeses. *Journal of Agricultural Food Chemistry*, 37:75–81.

Haider, M., and Haider, S. Q. (1984). Assessment of protein-calorie malnutrition. *Clin. Chem.*, 30:286–99.

Halliwell, B. (1988). Albumin—an important extracellular antioxidant? *Biochem. Pharmacol.*, 37:569–71.

Halliwell, B., and Gutteridge, J. M. C. (1986). Oxygen free radicals and iron in relation to biology and medicine: some problems and concepts. *Arch. Biochem. Biophys.*, 246:501–14.

Hasuo, Y., Ueda, K., Fujii, I. et al. (1986). Changes in blood chemical constituents and death in the aged population: the Hisayama study. *Japanese Journal of Geriatrics*, 23:65–72.

Hata, H., Matsuzaki, H., and Takatsuki, K. (1990). Autocrine growth by two cytokines, interleukin 6 and tumor necrosis factor alpha in the myeloma cell line KHM-1A. *Acta Haematol. (Basel)*, 83:133–36.

Helman, N., and Rubinstein, L. S. (1975). The effects of age, sex and smoking on erythrocytes. *Am. J. Clin. Pathol.*, 63:35–44.

Hoidal, J. R., and Niewoehner, D. E. (1982). Lung phagocyte recruitment and metabolic alterations induced by cigarette smoke in humans and hamsters. *Am. Rev. Respir. Dis.*, 126:548–52.

Ichimaru, M., Ishimaru, T., Mikami, M., and Matsunaga, M. (1982). Multiple myeloma among atomic bomb survivors in Hiroshima and Nagasaki by marrow dose, 1950–1976. *J. Natl. Cancer. Inst.*, 69:323–28.

Imlay, J. A., and Linn, S. (1988). DNA damage and oxygen radical toxicity. *Science*, 240:1302–09.

Jacobs, A. (1985). Ferritin: an interim review. *Current Topics in Hematology*, 5:25–62.

Junien, C., Turleau, C., deGrouchy, J., Said, R., Rethore, M. O., Tenconi, R., et al. (1980). Regional assignment of catalase gene to band 11p13. Association with the anirida Wilms' tumor-gonadoblastoma complex. *Ann. Genet.*, 3:156–58.

Kasama, T., Kobayashi, K., Fukushima, T., Tabata, M., Ohno, I., Negishi, M., et al. (1989). Production of interleukin-1 like factor from human peripheral monocytes and polymorphonuclear leukocytes by superoxide anion: The role of interleukin 1 and reactive oxygen species in inflamed sites. *Clin. Immunol. Immunopathol.* 53(3):439–48.

Kato, H., Mayumi, M., Nishioka, K., and Hamilton, H. (1983). The relationship of HB surface antigen and antibody to atomic bomb radiation in the adult health study sample, 1975-77. *Am. J. Epidemiol.*, 117:610–20.

Kawano, M., Hirano, T., and Matsuda, T. (1988). Autocrine generation and requirement of BSF-2/IL-6 for human multiple myeloma. *Nature*, 332:83–5.

Kono, Y., and Fridovich, I. (1982). Superoxide radical inhibits catalase. *J. Biol. Chem.*, 257:5751–54.

Konopka, K., and Neilands, J. B. (1984). Effect of serum albumin on siderophore-mediated utilization of transferrin iron. *Biochemistry*, 23:2122–27.

Kravitz, K., Skolnick, M., Cannings, C. et al. (1979). Genetic linkage between hereditary hemochromatosis and HLA. *Am. J. Hum. Genet.*, 31:601–19.

Kumagai, E., Tanaka, R., Kumagai, T., Higashida, Y., Onomichi, M., Katsuki, T. et al. (1985). Effects of low level radiation doses on man. (1) With reference to Epstein-Barr virus-associated antibody titers. *Japanese Journal of Clinical Pathology*, 33:1451–56.

Larramendy, M., Mello-Filho, A. C., Martins, E. A., et al. (1987). Iron-mediated induction of sister-chromatid exchanges by hydrogen peroxide and superoxide anion. *Mutat. Res.*, 178:57–63.

Lesko, S. A., Drocourt, J.-L., and Yang, S.-U. (1982). Deoxyribonucleic acid-protein and deoxyribonucleic acid interstrand cross-links induced in isolated chromatin by hydrogen peroxide and ferrous ethylenediaminetetraacetate chelates. *Biochemistry*, 21:5010–15.

Link, E. M. (1987). Comment, *Br. J. Cancer*, 55(Suppl):110–12.

Lipecka, K., Graboeska, B., Daniszewska, K., Domanski, T., and Cisowska, G. (1984). Correlation between the superoxide dismutase activity in lymphocytes and the yield of radiation-induced chromosome aberrations. *Studia Biophysica*, 100:211–17.

Lippman, S. M., and Meyskens, F. L. (1989). Retinoids for the prevention of cancer. In T. E. Moon and M. S. Micozzi, eds., *Nutrition and Cancer Prevention: Investigating the Role of Micronutrients*. New York: Marcel Dekkar pp. 243–71.

Loeb, L. A., James, E. A., Waltersdorph, A. M., and Klebanoff, S. J. (1988). Mutagenesis by autoxidation of iron with isolated DNA. *Proc. Natl. Acad. Sci. USA*, 85:3918–22.

Lunec, J. (1984). Free radical mediated aggregation of human IgG stimulates neutrophiles to generate superoxide radicals. *Agents Actions*, 15:37–9.

Marttila, R. J. (1988). Oxygen toxicity protecting enzymes in Parkinson's disease: Increase of superoxide dismutase-like activity in the substantia nigra and basal nucleus. *J. Neurol. Sci.*, 88:321–31.

Meijer, C., Mulder, N. H., Timmer-Bosscha, H. et al. (1987). Role of free radicals in an adriamycin-resistant human small cell lung cancer cell line. *Cancer Res.*, 47:4613–17.

Micozzi, M. S. (1989). Foods, micronutrients, and reduction of human cancer. In T. E. Moon and M. S. Micozzi, eds. *Nutrition and Cancer Prevention:*

Investigating the Role of Micronutrients. New York: Marcel Dekkar, pp. 213–241.
Micozzi, M. S., Beecher, G. R., Taylor, P. R. et al. (1990). Carotenoid analysis of selected raw and cooked foods associated with a lower risk for cancer. *J. Natl. Cancer Inst.*, 282–85.
Moon, T. E., and Micozzi, M. S., eds. (1989). *Nutrition and Cancer Prevention: Investigating the Role of Micronutrients*. New York: Marcel Dekker.
Nakamura, N., Kushiro, J., Sposto, R. et al. (1989). Is variation in human radiosensitivity real or artificial? A study by colony formation method using peripheral blood T-lymphocytes. RERF TR 15-89.
Nakamura, Y., Gindhart, T. D., Winterstein, D. et al: (1988). Early superoxide dismutase-sensitive event promotes neoplastic transformation in mouse epidermal JB6 cells. *Carcinogenesis*, 9:203–07.
Nakayama, T., Church, D. F., Pryor, W. A. (1989). Quantitative analysis of the hydrogen peroxide formed in aqueous cigarette tar extracts. *Free Radic. Biol. Med.*, 7:9–15.
Nathan, G., Arrick, B. A., Murray, H. W., Desantis, N. M., and Cohn, Z. H. (1988). Tumor cell anti-oxidant defence. Inhibition of the glutathine redox cycle enhances macrophage-mediated cytolysis. *J. Exp. Med.*, 135:766–82.
National Research Council. (1989). *Diet and Health: Implications for Reducing Chronic Disease Risk*. Washington, D.C.: National Academy Press.
Neriishi, K., Stram, D. O., Veath, M., Mizuno, S., and Akiba, S. (1989). The observed relationship between acute radiation sickness and subsequent cancer mortality among atomic bomb survivors in Hiroshima and Nagasaki. RERF TR 18-89.
Neriishi, K., Matsuo, T., Ishimaru, T., and Hosoda, Y. (1986). Radiation exposure and serum protein alpha and beta globulin fraction. *Nagasaki Medical Journal*, 61:223–28.
Nicotera, T. M., Notaro, J., Notaro, S., Schumer, J., and Sandberg, A. A. (1989). Elevated superoxide dismutase in Bloom's syndrome—A genetic condition of oxidative stress. *Cancer Res.*, 49:5239–43.
Niederau, C., Fischer, R., Sonnenberg, A. et al. (1985). Survival and causes of death in cirrhotic and noncirrhotic patients with primary hemochromatosis. *N. Eng. J. Med.*, 313:1256–62.
Nishigori, C., Miyachi, Y., Imamura, S., and Takebe, H. (1989). Reduced superoxide dismutase activity in xeroderma pigmentosum fibroblasts. *J. Invest. Dermatol.*, 93:506–10.
Nishimoto, N., Yoshizaki, K., and Tagoh, H. (1989). Elevation of serum interleukin 6 prior to acute phase proteins on the inflammation by surgical operation. *Clin. Immunol. Immunopathol.*, 50:399–401.
Omenn, G. S. (1988). A double-blind randomized trial with beta-carotene and retinol in persons at high risk of lung cancer due to occupational asbestos exposures and/or cigarette smoking. *Public Health Reviews*, 16:99–125.
Otsuka, F., Tarone, R. E., Seguin, I. R., and Robbins, J. H. (1985). Hypersensitivity to ionizing radiation in cultured cells from Down syndrome patients. *J. Neuro. Sci.*, 69:103–12.
Oya, Y., Tonomura, A., and Yamamoto, K. (1987). The biological activity of hydrogen peroxide. III. Induction of Epstein-Barr virus via indirect action, as compared with TPA and teleocidin. *Int. J. Cancer*, 40:69–73.

Ozaki, Y., Ohashi, T., and Kume S. (1987). Potentiation of neutrophile function by recombinant DNA-produced interleukin 1. *J. Leukoc. Biol.*, 42:621–27.

Pariza, M. W. (1988). Dietary fat and cancer risk. *Annu. Rev. Nutr.*, 8:167–83.

Peters, T. (1985). Serum albumin. *Adv. Protein Chem.*, 37:161–245.

Petkau, A., Chelack, W. S., and Pleskach, S. D. (1976). Protection of post-irradiated mice by superoxide dismutase. *Int. J. Radiat. Biol.*, 29(3):297–99.

Phillips, A., Shaper, A. G., and Whincup, P. H. (1989b). Association between serum albumin and mortality from cardiovascular disease, cancer, and other causes. *Lancet*, 2:1 434–36.

Richter, C., Park, J.-W., and Ames, B. N. (1988). Normal oxidative damage to mitochondrial and nuclear DNA is extensive. *Proc. Natl. Acad. Sci. USA*, 85:6465–67.

Sabatier, L., Hoffschir, F., Achkar, W. A., Turleau, C., Grouchy, J., and Dutrillaux, B. (1989). The decrease of catalase or esterase D activity in patients with microdeletions of 11p or 13q does not increase their radiosensitivity. *Ann. Genet.*, 32:144–48.

Saggu, H., Cooksey, J., Dexter, F., Wells, F. R., Lees, A., Jenner, P. et al. (1989). A selective increase in particulate superoxide dismutase activity in Parkinsonian substantia nigra. *J. Neurochem.*, 53:692–97.

Samokyszyn, V. M., Thomas, C. E., Reif, D. W., Saito, M., and Aust, S. D. (1988). Release of iron from ferritin and its role in oxygen radical toxicities. *Drug Metab. Rev.*, 19:283–303.

Sandstrom, B. E., Carlsson, J., and Marklund, S. L. (1989a). Selenite-induced variation in glutathione peroxidase activity of three mammalian cell lines: no effect on radiation-induced cell killing or DNA strand breakage. *Radiat. Res.*, 117:318–25.

Sandstrom, B. E., Grankvist, K., and Marklund, S. L. (1989). Selenite-induced increase in glutathione peroxidase activity protects human cells from hydrogen peroxide-induced DNA damage, but not from damage inflicted by ionizing radiation. *Int. J. Radiat. Biol.*, 56:837–41.

Sawada, H., Kodama, K., Shimizu, Y., and Kato, H. (1986). Adult health study report 6, Results of six examination cycles, 1968–80, Hiroshima and Nagasaki. RERF TR 3-86.

Scott, M. D., Meshnick, S. R., and Eaton, J. W. (1989). Superoxide dismutase amplifies organism sensitivity to ionizing radiation. *J. Biol. Chem.*, 264:2498–501.

Seidegard, J., DePierre, J. W., and Pero, R. W. (1985). Hereditary interindividual differences in the glutathione transferase activity towards trans-stilbene oxide in resting human mononuclear leukocytes are due to a particular isozyme(s). *Carcinogenesis*, 6:1211–16.

Selby, J. V., and Friedman, G. D. (1988). Epidemiological evidence of an association of body iron stores and risk of cancer. *Int. J. Cancer.* 41:677–82.

Sposto, R., Stram, D. O., and Awa, A. A. (1990). An investigation of random errors in the DS86 dosimetry using data on chromosome aberrations and severe epilation. RERF TR 7-90.

Stevens, G. N., Joiner, M. C., Joiner, B. et al. (1989). Role of glutathione peroxidase in the radiation response of mouse kidney. *Int. J. Radiat. Oncol. Biol. Phys.*, 16:1213–17.

Stevens, R. G., Kuvibidila, S., Kapps, M., Friedlaender, J. S., and Blumberg, B. S. (1983). Iron-binding proteins, hepatitis B virus, and mortality in the Solomon Islands. *Am. J. Epidemiol.*, 118:550–61.

Stevens, R. G., Beasley, R. P., and Blumberg, B. S. (1986). Iron-binding proteins and risk of cancer in Taiwan. *J. Natl. Cancer Inst.*, 76:605–10.

Stevens, R. G., Jones, D. Y., Micozzi, M. S., and Taylor, P. R. (1988). Body iron stores and the risk of cancer. *N. Engl. J. Med.*, 319:1047–52.

Stevens, R. G., and Kalkwarf, D. R. (1990). Iron, radiation, and cancer. *Environ. Health Perspect.*, 87:291–300.

Stevens, R. G., and Blumberg, B. S. (1990). Serum albumin and chronic disease mortality. *Lancet*, 335; 51.

Stevens, R. G. (1990). Iron and the risk of cancer. *Med. Oncol. Tumor Pharmacother.*, 7:177–81.

Stram, D. O., Akiba, S., Neriishi, K., Stevens, R. G., and Hosoda, Y. (1990). Smoking and serum proteins in atomic bomb survivors in Hiroshima. RERF TR 3-89 and *Am. J. Epidemiol.*, 131:1038–45.

Sullivan, J. L. (1982). Iron, aspirin, and heart disease risk. *JAMA*, 247:751.

Sullivan, J. L. (1989). The iron paradigm of ischemic heart disease. *Am. Heart J.*, 117:1177–88.

The Surgeon General: *Nutrition and Health.* (1988). U.S. Department of Health and Human Services, Public Health Service, DHHS (PHS) Publication number 88-50210, U.S. Government Printing Office, Washington, D.C.

Svejgaard, A., Platz, P., and Ryder, L. P. (1983). HLA and disease 1982—a survey. *Immunol. Rev.* 70:193–218.

Van Rensburg, C. E. J., Theron, A., Richards, G. A. et al. (1989). Investigation of the relationship between plasma levels of ascorbate, vitamin E, β-carotene and the frequency of sister-chromatid exchange and release of reactive oxidants by blood leukocytes from cigarette smokers. *Mutat. Res.*, 215:167–72.

Vuillume, M., Calvayrac, R., Best-Belpomme, M. et al. (1986). Deficiency in the catalase activity of xeroderma pigmentosum cell and simian virus 40-transformed human cell extracts. *Cancer Res.*, 46:538–44.

Ward, J. F., Blakely, W. F., and Joner, E. I. (1985). Mammalian cells are not killed by DNA single-strand breaks caused by hydroxyl radicals from hydrogen peroxide. *Radiat. Res.*, 103:383–92.

Weinberg, E. D. (1984). Iron withholding: a defense against infection and neoplasia. *Physiol. Rev.*, 64:65–102.

Whiting, R. F., Wei, L., and Stich, H. F. (1981). Chromosome-damaging activity of ferritin and its relation to chelation and reduction of iron. *Cancer Res.*, 41:1628–36.

Wu, W.-H., Meydani, M., Meydani, S. N. et al. (1990). Effect of dietary iron overload on lipid peroxidation, prostaglandin synthesis and lymphocyte proliferation in young and old rats. *J. Nutr.*, 120:280–89.

Yada, B., Goto, Y., Sato, K. et al. (1985). Ultra-weak chemiluminescence of smoker's blood. *Arch. Environ. Health*, 40:148–50.

Yamakido, M., Akiyama, M., Dock, D. S. et al. (1983). T and B cells and PHA response of peripheral lymphocytes among atomic bomb survivors. *Radiat. Res.*, 93:572–80.

Zoschke, D. C., and Staite, N. D. (1987). Suppression of human lymphocyte proliferation by activated neutrophiles or H_2O_2: Surviving cells have an altered T helper/T suppressor ratio and increased resistance to secondary oxidant exposure. *Clin. Immunol. Immunopathol.*, 42:160–70.

Zurlo, M. G., DeStefano, P., Borgna-Pignatti, C. et al. (1989). Survival and causes of death in thalassemia major. *Lancet*, 2:27–30.

8

Modulation of Carcinogenesis by Antioxidants

THOMAS W. KENSLER AND KATHRYN Z. GUYTON

It was first observed in the early part of this century that carcinogenesis could be modified by discrete chemical agents (see Talalay et al. 1987). The field of cancer chemoprotection did not begin to receive significant attention, however, until the 1960s when Wattenberg (reviewed in Wattenberg 1980; Wattenberg 1985) demonstrated that dietary antioxidants can protect against tumor formation. The experimental observations that seemingly innocuous food additives could dramatically protect against diverse carcinogens sparked the development of chemoprotection as a viable strategy for the reduction of human cancers. Additional interest in the relation between diet and cancer in humans has been stimulated by the observed differences in cancer incidence rates among countries. Recent epidemiologic evidence suggests that a decreased incidence of and mortality from cancer are correlated with the consumption of antioxidants (Bertram et al. 1987). The association of dietary intake of the antioxidants selenium, β-carotene, and vitamins C and E with reduced human cancer incidence is reviewed in Chapter 7.

In the past decade numerous animal studies have documented the protection against a diverse array of chemical carcinogens that can be accomplished with an equally diverse group of inhibitory compounds. In addition to phenolic food antioxidants such as butylated hydroxyanisole (BHA), butylated hydroxytoluene (BHT), and the tocopherols, important chemical classes of chemoprotectors include indoles, organic isothiocyanates, coumarins, flavones, dithiocarbamates, retinoids, and dithiolethiones. Although protection is achieved against chemically unrelated carcinogens, it appears that these chemoprotectors function in three basic ways: as modulators of the metabolic processing of carcinogens; as nucleophilic trapping agents that either block the nonenzymatic formation of carcinogens or intercept ultimate carcinogens; and as scavengers of free radicals produced from endo- and xenobiotic sources. Antioxidants, in their roles as chemoprotectors, may act through all of these mechanisms. However, experimental evidence to date suggests that they act principally by upsetting the balance between metabolic activation and detoxication of carcinogens and by scavenging the free radical species produced during the later stages of the neoplastic process.

Several factors presently limit the application of antioxidants as potential chemoprotective agents in humans. Issues of safety and toxicity contraindicate the long-term, high-dose administration of antioxidants to healthy people in chemoprotection interventions. Several synthetic antioxidants are also carcinogenic in their own right, and it is not clear that even natural antioxidants can be taken at pharmacologic levels with impunity. Furthermore, improved biochemical and biologic markers must be developed that can serve both to identify populations at risk and to provide intermediate end points that may later predict reduction in cancer incidence rates. This chapter includes a brief discussion of how antioxidants function followed by a discussion of mechanisms of carcinogenesis and carcinogen activation with a focus on how antioxidants interplay in these processes. Through a fuller appreciation of the underlying mechanisms of carcinogenesis and anticarcinogenesis, it may be possible to separate the chemoprotective and toxicologic properties of antioxidants and to enhance their application to human populations.

AUTOXIDATION REACTIONS AND ANTIOXIDANT ACTION

Antioxidants have had a long-standing commercial role as antagonists of autoxidation processes. Air-induced oxidation, or autoxidation, is responsible for the deterioration of a variety of inorganic and organic materials: a familiar consequence is rancidity in foods. More complex and less evident, however, are the effects on biologic systems leading to atherosclerosis, aging, and cancer. Autoxidation involves the processes of initiation, propagation, and termination of free radicals and is summarized in Figure 8-1. Antioxidants terminate free radical reactions principally by functioning as reducing agents (electron or H-atom donors) or peroxyl radical chain interrupters. Antioxidants therefore interact with highly reactive radical species to yield a second, less reactive free radical. This secondary radical is sufficiently stable to preferentially undergo chain termination rather than initiate further organic radical formation. Both synthetic and natural antioxidants typically include as part of their molecular structure an aromatic ring to delocalize the free electron of a radical and one or more hydroxyl groups to provide labile hydrogen atoms. Many of the natural antioxidants are derivatives or isomers of flavones, isoflavones, flavonols, catechins, eugenol, coumarin, tocopherols, cinnamic acid, carotenoids, phosphatides,

INITIATION

$$H_2R + X\bullet \longrightarrow HR\bullet + HX$$

PROPAGATION

$$HR\bullet + O_2 \longrightarrow HROO\bullet$$

$$HROO\bullet + H_2R \longrightarrow HROOH + HR\bullet$$

TERMINATION

$$2\ HROO\bullet \longrightarrow R{=}O + HRCH + O_2$$

$$HROO\bullet + AH \longrightarrow HROOH + A\bullet$$

Figure 8–1 The processes of initiation, propagation, and termination of free radicals. X• = free radical initiator; H_2R = active methylene group on fatty acid molecule; AH = phenolic antioxidant.

and polyfunctional organic acids. Some antioxidants commonly used in anticarcinogenesis studies are shown in Figure 8-2.

CANCER CHEMOPROTECTION BY ANTIOXIDANTS

Multistage Carcinogenesis: Electrophile and Oxidant Stress

Experimental and epidemiologic studies have revealed many causative agents in carcinogenesis and have established that the induction of cancer by chemicals involves stages of cellular evolution from normal, through preneoplastic and premalignant cells, to highly malignant neoplasia. In several model systems, the stages of initiation, promotion, and progression can be operationally defined through the use of discrete chemical agents. Initiation typically requires only a single exposure to a carcinogen at a subthreshold dose and is generally believed to result from interaction of the ultimate carcinogenic species with the cellular genome. Promotion, which follows initiation, requires repeated exposures and is a phenomenon of gene activation in which the latent phenotype of the initiated cell becomes expressed. A final stage, progression, involves the conversion of benign tumors into malignant neoplasms. As is discussed in more detail below, substantial evidence suggests that free radicals and the induction of a prooxidant state play prominent roles in the stages of tumor promotion and progression. Overall, this multistage construct is expedient for the experimentalist and has led to much of our knowledge about the roles of antioxidants in carcinogenesis. It must be recognized, however, that human exposures to antioxidants and carcinogens tend to be concurrent and continuous and occur at lower concentrations than used in animal studies.

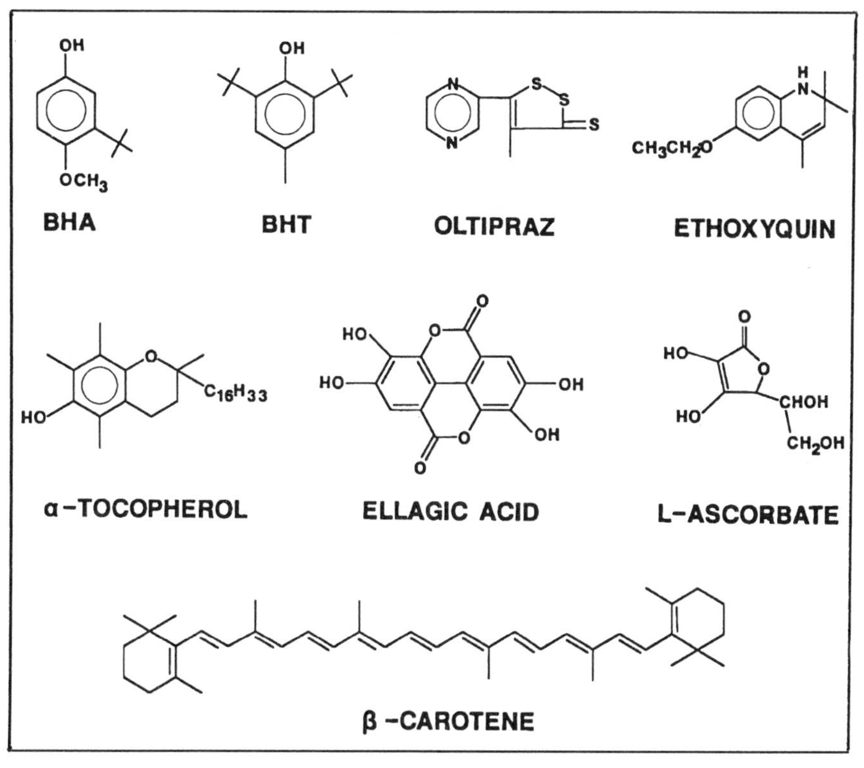

Figure 8–2 Antioxidants commonly used in anticarcinogenesis studies. BHA = butylated hydroxyanisole; BHT = butylated hydroxytoluene.

These factors make direct extrapolations of experimental effects and mechanisms to humans somewhat problematic.

A key component in understanding the initial events of carcinogens was the recognition by the Millers that many chemical carcinogens are not chemically reactive per se, but must undergo metabolic activation to form an electrophilic reactant (reviewed in Miller, 1970). These reactive species can interact with nucleophilic groups in DNA to induce point mutations and other genetic lesions, often leading to proto-oncogene activation. The importance of metabolic activation in carcinogenesis is highlighted by the fact that target organ specificities and even species susceptibilities can be determined through the presence or absence of metabolic activation pathways. The metabolism of chemicals to proximate carcinogens often involves an initial two-electron oxidation to a hydroxylated or epoxidated product and is typically catalyzed by the cytochrome P450 system. It is becoming increasingly apparent, however, that one-electron oxidations or reductions of procarcinogens yielding a radical intermediate having an odd or spin unpaired electron in its outer orbital may also play critical roles in the activation of carcinogens to DNA-damaging species. Collectively, the enzymes that catalyze the formation of these reactive intermediates are termed phase I enzymes.

A basic consideration that has emerged from investigations on the bioactivation of chemicals to reactive intermediates in vivo is that cells have a variety of enzymatic and nonenzymatic mechanisms that protect against damage by electrophilic and free radical metabolites. Thus, the amount of ultimate carcinogen available for interaction with its target represents, in part, a balance between competing activating and detoxifying reactions. A number of enzymes transfer or conjugate various endogenous substrates, such as glutathione, glucuronide, and sulfate, to the products of phase I metabolism. These phase II reactions, which often add large polar molecules to the primary metabolite, generally limit further biotransformation by enhancing elimination, thereby leading to detoxication. Examples of phase II enzymes include epoxide hydrolases, glutathione and glucuronyl transferases, and NAD(P)H:quinone reductase. Although a chemical may be able to undergo metabolism to an intermediate that can function as an initiator of carcinogenesis, the likelihood of this occurring is determined by both the absolute activities of and the relative balance between activating and detoxifying enzyme systems. Although this balance is under genetic control, it is easily modulated by exposure to drugs or other xenobiotics, nutritional status, age, and hormones. Antioxidants can profoundly modulate the constitutive metabolic balance between activation and inactivation of carcinogens through their actions on both phase I and II enzymes.

Inhibition of Carcinogen Activation by Antioxidants

Free radicals are also able to mediate the activation of carcinogens to electrophilic moieties. For example, some radicals such as peroxyl radicals (ROO·) may have an indirect role in carcinogenesis as mediators of hydroperoxide (ROOH)-dependent oxidations of carcinogens. There are a number of peroxidases that may participate in this route of metabolic activation by co-oxidation of procarcinogens including lactoperoxidase, uterine peroxidase, thyroid peroxidase, myeloperoxidase of polymorphonuclear leukocytes, and the hydroperoxidase component of prostaglandin synthase. As examples, aflatoxin B_1, aromatic amines, and polycyclic aromatic hydrocarbon dihydrodiols are oxidized to mutagenic derivatives during prostaglandin biosynthesis or by leukocytes undergoing an oxidative respiratory burst (Marnett 1987; Trush et al. 1985). In particular, peroxyl radicals contribute to the oxidation of (+)-benzo[a]pyrene-7,8-diol in crude tissue preparation, cultured fibroblasts, hamster tracheal explants, and primary cultures of murine keratinocytes and in mouse skin in vivo (Pruess-Schwartz et al. 1989). In several of these settings, the phenolic antioxidants BHA and BHT block this peroxyl radical-mediated activation of benzo[a]pyrene-7,8-diol. Concordantly, co-administration of either of these antioxidants with benzo[a]pyrene blocks the initiating activity of this hydrocarbon in mouse skin. Thus, antioxidants and other free radical scavengers block the formation of mutagenic metabolites by these systems. Although the chemistry of peroxyl radical-mediated activation by procarcinogens has been well characterized in in vitro systems, the relevance of these pathways to carcinogen activation in intact organisms remains unclear.

Antioxidants can also modulate the activities and levels of expression of cytochrome P450s. For example, Cha and Heine (1982) observed that both BHA and BHT enhanced the relative activity of aniline ring hydroxylation but decreased the relative benzo[a]pyrene mono-oxygenase activities in microsomes prepared from the livers of treated mice. Yang and colleagues (1981) also described inhibitory effects of BHA and BHT on aryl hydrocarbon hydroxylase activity, as measured by benzo[a]pyrene hydroxylation, in liver, lung, and skin of mice and rats. Further, dietary ethoxyquin increased microsomal cytochrome P450 activities leading to enhanced formation of presumed detoxication metabolites of aflatoxin B_1, aflatoxins M_1 and Q_1 (Mandel et al. 1987). Relatively less of the active metabolite, aflatoxin-8,9-oxide, was generated, leading to an overall tilt in the metabolic balance toward detoxication. Metabolism studies with hepatocytes isolated from rats previously fed BHT also demonstated an enhanced formation of metabolites of aflatoxin (Salocks et al. 1984). Pretreatment with BHT led to both enhanced oxidation of aflatoxin and accelerated aflatoxin conjugation. By contrast, if BHA is added concurrently with aflatoxin to either liver microsomes or primary hepatocytes, the oxidative metabolism of aflatoxin B_1 is inhibited by 90% (Ch'ih et al. 1989). This effect is apparently related to the noncompetitive inhibition of P450 activity by BHA. Collectively, these observations demonstrate that antioxidants can modify phase I reactions in a variety of ways, depending on both the concentration of antioxidant and the temporal relationship between the exposures to antioxidant and carcinogen.

Induction of Electrophile Detoxication Enzymes by Antioxidants: Consequences and Mechanisms

Most cancer chemoprotective studies utilize experimental designs in which anticarcinogens are administered before and concurrent with exposure to the carcinogen. Chemoprotective agents typically are given either in the diet or as discrete single administrations at some specific time interval before carcinogen treatment. In general, such experimental protocols examine the involvement of chemoprotective agents in the initiation phase of carcinogenesis. Table 8-1 lists several model systems in which protective effects for four antioxidants, BHA, BHT, ethoxyquin, and oltipraz, have been demonstrated when they are administered either before or simultaneously with the carcinogen. Although it has been known for many years that antioxidants exert an anticarcinogenic effect when given simultaneously with a carcinogen, there have been few experiments designed to investigate the mechanisms of such protective actions. One of the earliest studies to implicate a role for the induction of phase II enzymes, particularly

Table 8–1 Inhibition of carcinogen-induced neoplasia by BHA, BHT, ethoxyquin, or oltipraz[a]

Target Organ	Carcinogen	Antioxidant	Species	Reference
Forestomach	BP	BHA,BHT,EQ,OLT	Mouse	Wattenberg (1972b); Wattenberg and Bueding (1986)
	DEN	BHT,OLT	Mouse	Wattenberg and Bueding (1986); Clapp et al. (1978)
	DMBA	BHA,BHT,EQ	Mouse	Wattenberg and Lam (1980)
Colon	MAM acetate	BHA	Mouse	Reddy and Maeura (1984)
	DMH	BHT	Mouse	Clapp et al. (1979)
	Azoxymethane	BHT	Rat	Zedeck et al. (1972)
Liver	2-AAF	BHT	Rat	Williams et al. (1983); Ulland et al. (1973)
	N-OH-AAF	BHT	Rat	Ulland et al. (1973)
	AFB_1	BHA,BHT,EQ,OLT	Rat	Williams et al. (1986); Cabral and Neal (1983); Roebuck et al. (1991)
	3′-Me-DAB	BHT	Rat	Daoud and Griffin (1980)
Lung	BP	BHA,EQ,OLT	Mouse	Wattenberg and Bueding (1986); Wattenberg (1973)
	DEN	BHA,EQ,OLT	Mouse	Wattenberg and Bueding (1986); Wattenberg (1972a)
	DMN	BHA	Mouse	Chung et al. (1986)
	DMBA	BHA	Mouse	Wattenberg and Bueding (1986)
	Uracil mustard	OLT	Mouse	Wattenberg and Bueding (1986)
	Urethane	BHA,BHT	Mouse	Wattenberg (1973); Witschi and Lock (1979)
	MNU	OLT	Hamster	Boone et al. (1990)
Mammary gland	DMBA	BHA,BHT,EQ,OLT	Rat	McCormick et al. (1984)
Pancreas	BOP	BHA	Hamster	Mizumoto et al. (1989)
Skin	DMBA	BHA,BHT	Mouse	Slaga and Bracken (1977)

[a]2-AAF = 2-acetylaminofluorene; N-OH-AAF = N-hydroxy-N-2-fluorenylacetamide; AFB_1 = aflatoxin B_1; BHA = butylated hydroxyanisole; BHT = butylated hydroxytoluene; BP = benzo[a]pyrene; BOP = N-nitrosobis(2-oxopropyl)amine; DEN = diethylnitrosamine; DMN = dimethylnitrosamine; DMBA = 7,12-dimethylbenz[a]anthracene; DMH = 1,2-dimethylhydrazine; EQ = ethoxyquin; MAM = methylazoxymethanol; 3′-Me-DAB = 3′-methyl-4-dimethylaminoazobenzene; MNU = methylnitrosourea; OLT = oltipraz [5-(2-pyrazinyl)-4-methyl-1,2-dithiole-3-thione].

glutathione transferases, in the protective actions of antioxidants was that of Talalay and coworkers (Benson et al. 1978). They showed that liver cytosols from BHA- or ethoxyquin-fed rats or mice exhibited much higher glutathione transferase activities and that cytosols prepared from the livers of these rodents inhibited the mutagenic activity in urine from mice treated with benzo[a]pyrene. Subsequent studies demonstrated that dietary administration of antioxidants increased glutathione transferse activity in extrahepatic tissues such as lung, stomach, small intestine, and kidney (Benson et al. 1979). In fact, several laboratories now use the induction of phase II enzymes like glutathione transferases and quinone reductase to guide the isolation and purification of new classes of naturally occurring anticarcinogens.

Experimental hepatocarcinogenesis in rodents can be inhibited by a number of antioxidants and is particularly suited for mechanistic studies. A brief discussion of the impact of antioxidants on aflatoxin-induced liver cancers will serve to illustrate some of the enzyme-inducing and anticarcinogenic mechanisms of antioxidants. Aflatoxin is a potent hepatotoxin and carcinogen in a wide variety of animals and is linked epidemiologically with a high incidence of primary hepatocellular carcinoma in humans. A number of studies in Africa and Southeast Asia have demonstrated positive associations between aflatoxin ingestion and liver cancer incidence in humans (Groopman et al. 1988). Over the past several years it has been demonstrated that feeding BHA, BHT, ethoxyquin, or oltipraz during the period of aflatoxin exposure dramatically reduces the incidence of hepatocellular carcinomas in the rat (see Table 8-1). These protective actions are thought to result primarily from an altered balance between the activation and detoxication of aflatoxin in the hepatocyte. Aflatoxin undergoes metabolic activation to a highly reactive 8,9-epoxide, which can bind to macromolecules, particularly, DNA. This epoxide can also be conjugated to glutathione through the actions of glutathione transferases, yielding a detoxication product. In the rat, the induction of glutathione transferases through transcriptional activation is a prominent biochemical effect of dietary antioxidant treatment (Davidson et al. 1990). Correspondingly, aflatoxin treatment of rats maintained on antioxidant-supplemented diets results in large increases in the biliary elimination of aflatoxin-glutathione conjugates and in greatly diminished levels of aflatoxin modification of hepatic DNA following either single or multiple exposures to this carcinogen (Kensler et al. 1987). Overall, a striking correlation exists between the degree of induction of hepatic glutathione transferases by structurally distinct antioxidants and the degree of chemoprotection as judged by reduced levels of the major aflatoxin-DNA adduct, aflatoxin-N^7-guanine, in rat liver DNA (Kensler et al. 1985).

The molecular mechanisms regulating the transcriptional activation of phase II enzymes by antioxidants have also been investigated. Prochaska and Talalay (1988) demonstrated the requirement for functional aryl hydrocarbon (Ah) receptors in the induction of quinone reductase activity by planar aromatic hydrocarbons using variant mouse hepatoma cells that were defective in number or translocation of Ah receptors into the nucleus. However, the mode of action of agents such as the polycyclic aromatic hydrocarbons, which mediate the induction of both phase I and II enzymes and are thus termed "bifunctional" inducers, is in fact distinct from that of the "monofunctional" inducers, which induce only phase II enzymes. In contrast to the bifunctional inducers, monofunctional inducers, such as the antioxidants *tert*-butyl hydroquinone or 1,2-dithiole-3-thione, are effective inducers of quinone reductase in receptor-deficient variants of the hepatoma cells. Thus, not all inducers of phase II enzymes mediate their effect through the Ah receptor. Indeed, in earlier studies, Prochaska and colleagues (1985) had evaluated the induction of glutathione transferase activity by monomeric and dimeric analogues of BHA and found that there was no strict structural specificity in the induction mechanisms. These studies implicated a chemical rather than a receptor-mediated signaling system in the control of enzyme induction. More recently, Talalay's group identified a common chemical signal by which many monofunctional inducers may mediate the induction of quinone reductase and glutathione transferases. In general, these inducers are Michael reaction receptors characterized by olefinic bonds that are rendered electrophilic by conjugation with electron-withdrawing substituents; the potency of inducers parallels their efficiency in Michael reactions (Talalay et al. 1988). Most inducers also appear to be substrates for glutathione transferases.

The relevance of these mechanisms to chemoprotection is highlighted by the studies of Seidegård and coworkers (1986) who suggest that deficiencies in the levels of expression of glutathione transferases in humans may be important determinants for susceptibility to lung cancer. Thus, the modulation of transferase expression by inducers such as antioxidants may offer a realistic strategy for cancer chemoprotection. However, the efficacy in humans of agents identified in rodent bioassays as functional inducers of transferases or other phase II enzymes remains to be established. Nonetheless, the development of antioxidant enzyme inducers as chemoprotective agents in humans will be greatly facilitated by increased understanding of the roles of these enzymes in carcinogenesis and of the mechanisms that regulate their expression.

Antioxidants as Nucleophilic Trapping Agents

Several naturally occurring antioxidants inhibit carcinogenesis through mechanisms entailing the prevention of reactive intermediate formation or the inhibition of the interaction of the ultimate carcinogen with nucleophilic targets. The abilities of ascorbic acid (vitamin C) and the polyphenol, ellagic acid, to inhibit carcinogenesis through these mechanisms provide examples of these processes.

Nitrosamines can produce tumors in the lung, liver, and gastrointestinal tract in experimental animals and are implicated in the etiology of gastric cancers in humans. The precursors of nitrosamines (nitrite and amines or amides) are present in many foods, air pollution, and cigarette smoke. The ingestion or inhalation of these precursor compounds can lead to the elaboration of reactive nitrosamines in the appropriate environment. Ascorbic acid has exhibited the ability to inhibit the formation of these reactive electrophilic species and to prevent nitrosamine-induced tumor formation in several model systems. For example, Mirvish and colleagues (1972) demonstrated that ascorbic acid could block nitrosamine formation in vitro. Further, feeding of ascorbic acid has been shown to prevent hepatotoxicity in rats and carcinogenicity in mice fed aminopyrine and sodium nitrate (Kamm et al. 1973; Greenblatt 1973). The inhibitory effects of ascorbic acid are believed to be achieved through competition with susceptible amines for the available nitrosating species. Thus, ascorbic acid acts to inhibit carcinogenesis by preventing the development of reactive electrophiles from carcinogen precursors.

Ellagic acid, a natural phenol present in soft fruits and vegetables, has also been demonstrated to inhibit the mutagenicity of and DNA alkylation by N-nitroso compounds. This activity is attributed to the ability of ellagic acid to prevent the interaction of the carcinogenic metabolite with critical sites on the DNA, namely the O^6-position of guanine (Barch and Fox 1988), although an inhibitory effect on metabolism has been suggested as an additional mechanism (Mandal et al. 1988). In the former case, ellagic acid may be altering the conformation of DNA, thus preventing the interaction of electrophilic nitrosamines with their putative cellular target. Ellagic acid can also inhibit the mutagenicity of benzo[a]pyrene 7,8-diol-9,10-epoxide-II in bacterial and mammalian cells in culture (Wood et al. 1982). In aqueous solution, ellagic acid forms covalent ether adducts with the diol epoxide, thereby enhancing the removal of the carcinogenic electrophile; it is primarily through this mechanism that ellagic acid exerts its antimutagenic effects. Similarly, the inhibitory effects of ellagic acid were noted on aflatoxin B_1 mutagenesis in *Salmonella typhimurium* and DNA damage in cultured rat and human tissues, again through the formation of noncritical adducts with aflatoxin B_1 epoxide (Mandal et al. 1987).

From a pharmacokinetic point of view it may be difficult to maintain effective concentrations of nucleophilic trapping agents in distal tissues throughout periods of carcinogen exposure. However, Wattenberg and colleagues (1987) have explored with some success the utility of aromatic thiols as electrophile trapping agents within the lumen of the gastrointestinal tract. These thiols, particularly if retained in the gastrointestinal tract, might be effective scavengers of electrophiles consumed in the diet or eliminated in the bile or both, following activation of procarcinogens in the liver.

Free Radicals in Tumor Promotion and Progression

Tumor promotion is a stage in carcinogenesis that encompasses events involved in the development of a benign neoplasia from initiated but preneoplastic tissue. This stage has been characterized by the use of chemical agents that are not by themselves carcinogenic, but that can modulate phenotypic expression in both initiated and noninitiated cells. In some circumstances, tumor promoters may modulate gene expression resulting in the proliferation rather than differentiation of initiated cells. Coordinately, selection and clonal expansion of initiated cells may be facilitated when tumor promoters accelerate terminal differentiation of surrounding noninitiated cells or when they are particularly toxic to them. Although tumor promoters are considered to be epigenetic in action, their effects are ultimately on the genome. Several lines of evidence suggest a role for free radicals in the process of tumor promotion. Although detailed reviews of the involvement of free radicals in tumor promotion can be found elsewhere (Cerutti 1985; Kensler and Taffe 1986), the experimental support can be summarized briefly. Firstly, reactive oxygen generating systems are able to mimic some of the in vitro actions of tumor promoters (Seed et al. 1987). Tumor promoters are also known to stimulate the endogenous production of oxygen radicals in several cell types. Further, tumor promoters provoke a rapid and sustained decrease in cellular antioxidant defenses, including superoxide dismutase (SOD), catalase, and glutathione peroxidase activities. The most direct evidence to date that free radicals are involved in tumor promotion and progression is the observation that free radical-generating compounds exhibit these activities (Slaga et al. 1981; O'Connell et al. 1986). Unfortunately, few investigators have sought to characterize the actual free radical species formed in target tissues, their routes of metabolic activation, and most importantly, their critical molecular targets. Much evidence for the role of free radicals in tumorigenesis comes from the frank inhibition of tumor promotion or progression or both by SOD-mimicking agents, the synthetic antioxidants BHA and BHT, or the naturally occurring antioxidants, α-tocopherol and glutathione (for reviews see: Kensler and Trush 1985; Perchellet and Perchellet 1989). It must be emphasized, though, that inhibition of a biologic process by antioxidants is not a sine qua non for free radical involvement in that process as is exemplified by the mechanisms of anticarcinogenesis described previously. The types of free radicals that may be involved in tumor promotion and progression are illustrated in Figure 8-3.

Antioxidants have also been shown to inhibit many of the biochemical changes that are linked to tumor promotion. For example, α-tocopherol prevented the inhibition of intracellular communication produced in response to liver tumor promoters or the generation of hydrogen peroxide by glucose oxidase in cultured murine hepatocytes (Ruch and Klaunig 1986). Using a different marker for promotion, Kozumbo and Cerutti (1986) demonstrated the inhibitory effect of BHT on poly-(ADP)-ribose accumulation in mouse C3H10T½ cells treated with 12-O-tetradecanoylphorbol-13-acetate (TPA). Poly-(ADP)-ribosylation of nuclear proteins is associated with their post-translational modification and may provide a mechanism through which gene modification, such as occurs in tumor pro-

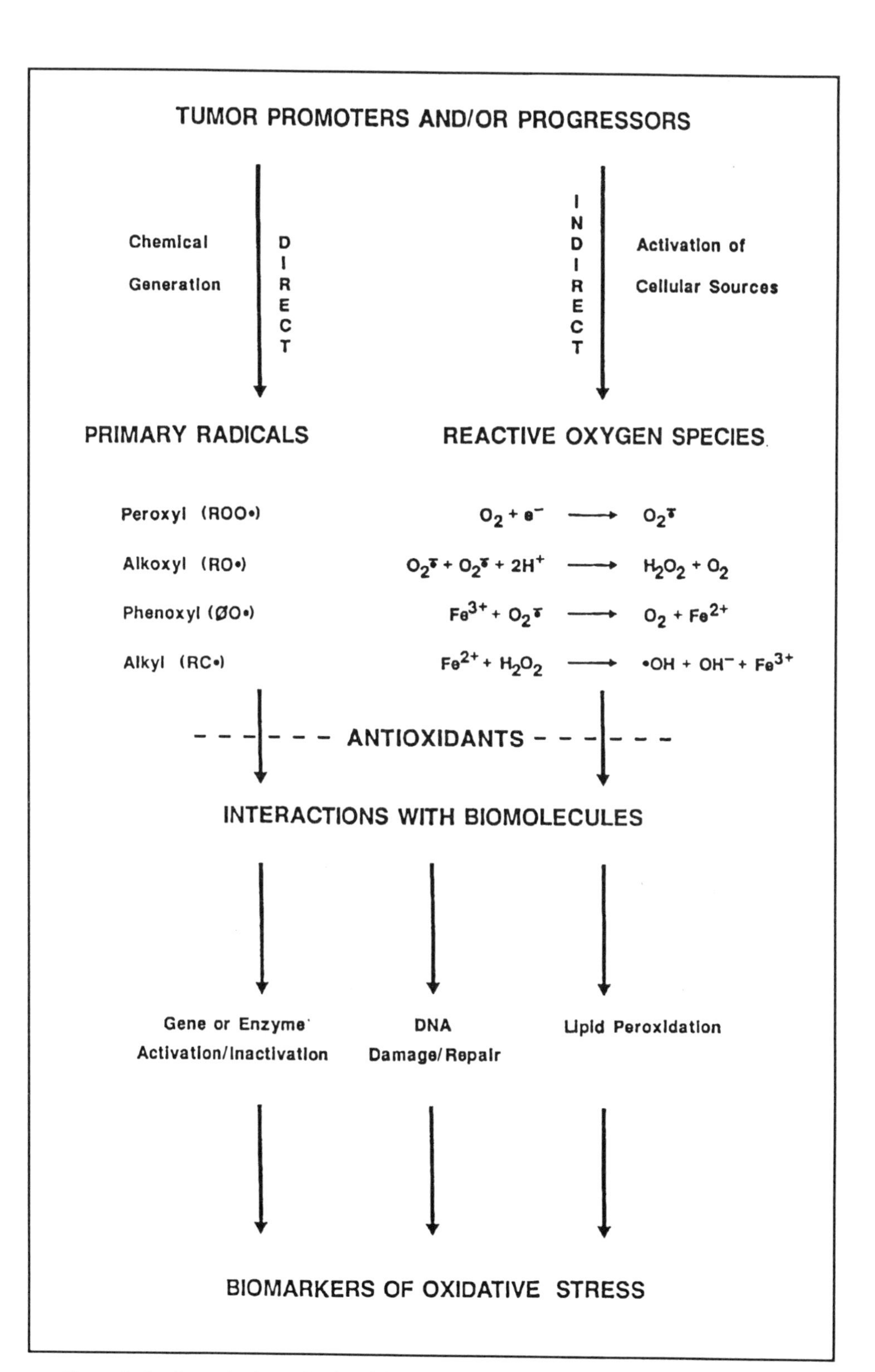

Figure 8–3 General scheme for the elaboration of free radical species by tumor promoters and progressors and the critical molecular targets involved in the development of a neoplastic state.

motion, may be regulated. The induction of ornithine decarboxylase (ODC) activity in response to tumor promoters has been widely used as a marker of the promotion stage of carcinogenesis. ODC catalyzes the initial and rate-limiting step in the biosynthesis of polyamines, whose production appears to be a necessary, although insufficient, event in tumor promotion. Several investigators have provided evidence that a variety of antioxidants can inhibit the induction of ODC in response to treatment of mouse epidermis with TPA or other tumor promoters, suggesting a central role for free radicals in promotion. A synthetic lipophilic antioxidant derivative of ascorbic acid, ascorbyl palmitate, substantially inhibited the induction of epidermal ODC activity after a single topical dose of TPA and reduced both DNA synthesis and mouse skin tumor promotion (Smart et al. 1987). Topical application of α-tocopherol, BHT, BHA, and several analogs of BHA, inhibited induction of ODC activity by TPA up to 80% in mouse epidermis (Kozumbo et al. 1983). BHA was ineffective when administered either 16 h before or 2 h after TPA, suggesting the BHA acts early and directly to inhibit TPA-induced increases in ODC activity. Perchellet and coworkers (1985) illustrated that D-α-tocopherol, glutathione, or the constituent amino acids of glutathione were able to inhibit TPA-induced ODC activity in vivo and in vitro by more than 50%. Furthermore, the inhibitory effects of these agents corresponded to their ability to decrease the fall in reduced/oxidized glutathione ratio stimulated by TPA in epidermal cells in vitro and to reduce the incidence of TPA-induced skin tumors in mouse skin. Other investigations concerning the impact of antioxidant treatment on the oxidant status of cells exposed to TPA include the study by Fischer and colleagues (1986), which demonstrated the ability of BHA, α-tocopherol, or other antioxidants to suppress TPA-stimulated chemiluminescence in murine epidermal cells. Cellular chemiluminescence is an indicator of excited-state molecules including reactive oxygen species. Kozumbo et al. (1985) showed that BHA and several of its analogs can diminish the chemiluminescent response of human polymorphonuclear leukocytes exposed to TPA. Taken together, these results are indicative of the diversity of promotion-related effects apparently mediated through free radical processes. It should be noted that, whereas the collective evidence strongly implicates a necessary role for free radicals in the actions of some classes of tumor promoters, it is likely that other signaling molecules may play even more important roles as mediators of tumor promoter action.

An additional component of carcinogenesis is the progression of a relatively benign lesion to a highly malignant, rapidly growing neoplasm. Although repetitive treatment of initiated mouse epidermis with phorbol esters is not particularly effective in the conversion of papillomas to carcinomas, subsequent treatment with initiating agents or peroxides enhances this progression to malignancy. This final stage may involve a second, discrete, heritable event that involves either further direct chemical interaction with DNA or the transposition of genetic material. The net result will be the irreversible transformation of one or more cells in the benign tumor into an autonomously growing cancer.

Free radical species are likely to participate in the progression stage of carcinogenesis as well. For example, in mouse skin, benzoyl peroxide, which is activated by keratinocytes to benzoyloxyl and phenyl radicals, is an effective enhancer of the conversion of benign papillomas to carcinomas. Rotstein and Slaga (1988) examined the effects of a variety of antioxidants as inhibitors of tumor progression in the murine multistage carcinogenesis model. Glutathione was the most effective inhibitor tested. It blocks progression of papillomas to carcinomas either in the presence or the absence of treatment with tumor-progressing agents like benzoyl peroxide. Additionally, repeated treatment with diethylmaleate, a glutathione-depleting agent, enhanced tumor progression. Thus, the level of the endogenous antioxidant glutathione appears to be an important determinant of susceptibility to tumor progression. Glutathione is also an effective inhibitor of DNA damage mediated by benzoyl peroxide, presumably by serving as an H-atom donor to detoxify benzoyloxyl radicals (Swauger et al. 1990).

ANTIOXIDANT INTERVENTIONS: SHORT-TERM MARKERS FOR EFFICACY

The major goal of chemoprotective interventions is to decrease the susceptibility of individuals to the consequences of carcinogen exposure. Because of the long latency period in most human cancers, it is impractical to design chemoprotection trials in which potential reduction of cancer incidence is the actual study end point. Thus, there is a growing interest in the development and characterization of intermediary biologic markers ("biomarkers") of exposure to chemicals and of the adverse health effects of such exposures (Wogan and Gorelick 1985; Henderson et al. 1989). A number of biomarkers for oxidative stress are beginning to be developed and validated. With such tools in hand, the evaluation of the potential modulation of human carcinogenesis by antioxidants will be greatly facilitated.

Strategies for developing biomarkers for exposure to as well as for response to oxidants are numerous and include measurement of specific metabolites or adducts formed by reactions of xenobiotics with macromolecules. Such measurements are made in body fluids, tissues, or exhaled air. Present analytical approaches are well suited for the measurements of DNA- or protein-adducts formed from bulky, electrophilic carcinogens. Measurement of DNA adducts appears to be particularly relevant as a dosimeter, because DNA adduct formation, if not repaired, is one of the critical initial steps in the neoplastic process. Because of the ephemeral nature of most free radicals, direct measurements of their levels in humans will be difficult. However, as reviewed by Kennedy and colleagues (1990), techniques utilizing spin trapping and electron paramagnetic resonance spectroscopy allow for the in vivo detection of free radicals in body fluids. Perhaps such studies can be extended to analyses for free-radical metabolites of antioxidants, such as phenoxyl radicals, as biomarkers of oxidative stress. The chapter by Linn in this volume describes a number of modifications to DNA bases following exposure to oxidants. Analogous to the situation with the bulky carcinogen adducts, these products of DNA damage induced by oxygen radicals may be useful biomarkers for oxidative stress. Dizdaroglu and Bergtold (1986) developed a sensitive method using gas chromatography-mass spectrometry with selected-ion-monitoring technique to quantify a large number of free radical-induced base products of DNA in a single sample of DNA. Utilizing instrumentation more amenable to routine biomonitoring, Ames and his colleagues developed several noninvasive assays for oxidative DNA damage in individuals through high performance liquid chromatography (HPLC) assays for thymine glycols (Cathcart et al. 1984) and 8-OH-guanine (Shigenaga et al. 1989) in human urine. These assays may allow for direct testing of hypotheses relating oxidant stress to carcinogenesis or other disease states and could be extended to monitoring chemoprotection interventions. Presumably, successful antioxidant interventions would reduce the levels of oxidized DNA products in the tissues and urine of exposed individuals. An additional possible reflection of oxidant stress might be the oxidation of lipids as measured in plasma lipoproteins (see Chapter 4).

As discussed in a previous monograph in this series (Bloom et al. 1989), a number of markers of susceptibility can be used to identify individuals at greater risk than the general population as a result of a genetic or other predisposition to the effects of exposure to environmental mutagens and carcinogens. In some instances, these predisposing factors relate to xenobiotic metabolism. Because one of the major effects of antioxidants is the induction of xenobiotic metabolism, methods developed for phenotyping the metabolic capacities of individuals could be effective markers of the pharmacologic efficacy of antioxidant treatment. For example, it may prove reasonable to monitor the changes in the levels of glutathione transferases or other phase II enzymes in peripheral lymphocytes or other accessible tissues following antioxidant intervention. However, unlike the approach of monitoring for antioxidant efficacy through diminution of carcinogen-DNA adduct/oxidized base levels, monitoring of enzyme-inductive effects in people requires greater surety in the underlying mechanisms of anticarcinogenic action by these agents.

ANTIOXIDANTS: A DOUBLE-EDGED SWORD?

Toxicity and Carcinogenicity of Antioxidants

Although the extensive and dramatic chemoprotective effects of BHA and BHT have engendered considerable hope that they may play a role in human cancer protection, a number of toxicologic properties of these phenolic antioxidants make this belief unlikely. This is not to suggest that the basic strategies of enzyme induction, nucleophile trapping, and free radical scavenging are not applicable to chemoprotection in man; however, candidate chemoprotective compounds, be they antioxidants or nonantioxidants, will require verification of lack of inherent short- and long-term toxicities. For this reason, some initial human chemoprevention trials have focused on naturally occurring, presumably less toxic, radical scavengers such as the tocopherols and carotenoids.

Despite the fact that administration of vitamin E resulting in pharmacologic blood levels appears to be tolerated with relative impunity, a number of toxic effects of vitamin E therapy have been described in humans. These effects include prolongation of plasma clotting time, inhibition of platelet prostaglandin synthesis and decreased platelet aggregation, and impaired immune function, as indicated by reduced bacteriocidal activity of leukocytes and depressed mitogen-induced transformation of lymphocytes. Further, there is little persuasive evidence to date for a compelling therapeutic role for vitamin E in humans. A toxic syndrome, hypervitaminosis A, results from excessive self-medication with vitamin A, food fads, or the use of high doses of oral retinoids for the therapy of skin disorders. Symptoms of chronic retinoid intoxication include dermatitis and skin desquamation, pathologic changes in the liver, and neurologic changes. Retinoids are also potent animal and human teratogens (Mandel and Cohn 1985). Thus, even the use of natural compounds in sustained prophylactic interventions will require close toxicologic scrutiny.

Because of their widespread use as food additives, the chronic toxicities of phenolic antioxidants such as BHT, BHA, and *tert*-butyl hydroquinone in animals have been studied extensively. Several excellent reviews have appeared on this topic in recent years (Kahl 1984; Ito et al. 1985; Ito and Hirose 1987). These antioxidants are not mutagenic as judged by a variety of short-term mutagenicity assays. However, BHA is carcinogenic in rats, hamsters, and mice. When high doses of BHA (1% or 2% in the diet) were fed to either rats or hamsters, a high incidence of papillomas and squamous cell carcinomas of the forestomach was observed. Experiments of shorter duration in animal species that, like humans, lack a forestomach (that is, guinea pig, monkey, and beagle dog) failed to demonstrate any hyperplastic or carcinogenic effect of BHA. Although the initial reports on the carcinogenicity of BHA raised significant concern over the continued use of BHA as a food additive, the knowledge that humans do not have forestomachs and that extremely high doses were required to demonstrate carcinogenicity has greatly tempered the initial reaction. At present, BHA remains the most abundantly used food antioxidant because no other effective antioxidants with a cleaner toxicologic profile have appeared as ready substitutes.

BHT which has more acute toxicities associated with exposure to it than does BHA, has a more ambiguous status as a possible human carcinogen.

A substantial number of one-generation bioassay studies in mice and rats indicate that BHT is not carcinogenic as a single agent. However, a recent report by Würtzen and Olsen (1986) indicated that a multigenerational exposure to BHT leads to hepatocellular carcinomas. In this study, F_0 rats of both sexes were fed graded concentrations of BHT and mated; F_1 rats were fed BHT at a concentration equivalent to a daily intake of 500 mg BHT/kg body weight for their lifetime. Dose-related increases in the numbers of hepatocellular adenomas and carcinomas were observed in male rats that were more than 2 years old. Furthermore, Inai and colleagues (1988) observed a dose-related increase in hepatocellular adenomas in male B6C3F_1 mice fed BHT at either 1% or 2% in the diet for 104 weeks. No effect on the incidence of hepatocellular carcinomas was observed nor was there any BHT-related effect on tumorigenicity in female mice. Taken together, these findings suggest that BHT is a relatively weak complete carcinogen in the liver.

Role of Free Radicals in Tumor Promotion by Phenolic Antioxidants

BHT has been reported to be a promoter in the intestine, colon, lung, thyroid, liver, and bladder in rats or mice or both when used in two-stage carcinogenesis protocols. Other commercial synthetic antioxidants, notably BHA and ethoxyquin, also act to enhance tumor formation in several of these rodent models. Unfortunately, in no case is the molecular mechanism underlying this phenomenon understood.

The best characterized system in which BHT acts to enhance tumor formation is the mouse lung. Weekly injections of BHT following treatment with the carcinogen urethane lead to significant increases in tumor multiplicity in A/J mice. Interestingly, if BHT was administered just before a single injection of urethane, fewer tumors were formed. Biotransformation of BHT was apparently required for both the prophylactic and tumor-enhancing effects, since pretreatment of mice with cedrene, which modulates xenobiotic metabolism, prevented both of these actions of BHT (Malkinson and Beer 1984). Recent evidence from Thompson and colleagues (1989) directly suggested that a metabolite of BHT may mediate its tumor-promoting activity in that a cytochrome P450-derived BHT metabolite, BHT-tBuOH, was several-fold more potent as an enhancer of urethane-induced tumorigenesis in the lung than was BHT. Further, Taffe and Kensler (1988) observed that another P450-dependent metabolite of BHT, BHT hydroperoxide, is an effective tumor promoter in mouse skin. BHT, which is not metabolized to the hydroperoxide by epidermal microsomes, is not a tumor promoter in this tissue. Electron paramagnetic resonance studies in isolated keratinocytes demonstrate the formation of a BHT-phenoxyl radical from BHT hydroperoxide. Thus, BHT may undergo metabolic activation to free radicals or other reactive intermediates (that is, quinone methides) in target tissues (Taffe et al. 1989; Guyton et al. 1991). The molecular mechanisms by which these reactive metabolites mediate the phenotypic changes induced by gene activation and cell proliferation are unclear. They may parallel the intracellular cascades utilized by growth factors, as is beginning to be described for reactive species derived from molecular oxygen. This view is discussed in detail in Chapter 6.

CONCLUSIONS

This chapter has focused on the paradoxical actions of antioxidants that might result in either chemoprotection or carcinogenesis. Effects of antioxidants that contribute to their actions as chemoprotectors include their ability to modulate carcinogen metabolism via effects on phase I and phase II enzyme activities; their role as nucleophilic trapping compounds that can inactivate electrophilic initiating agents; and their capacity as antioxidants to scavenge free radicals produced during the latter stages of carcinogenesis. Unfortunately, considerable evidence also supports a role for these same antioxidants as complete carcinogens or tumor promoters under other conditions. Because the effective treatment of many human cancers remains a difficult problem, the development of protective measures designed to reduce man's susceptibility to the actions of carcinogens is an important objective. Thus, future work should be directed toward a number of important areas:

1. Clarification of the roles of free radicals in the multiple stages of the carcinogenic process, particularly as related to the involvement of oxidants derived from chronic inflammation
2. Mechanism-based, structure-activity studies designed to dissociate the chemoprotective and carcinogenic properties of antioxidants with the hope of identifying antioxidant compounds with selective chemoprotective characteristics
3. Identification of high-risk populations appropriate for antioxidant chemoprotection interventions
4. Development and validation of short-term markers for assessing the efficacy of antioxidant interventions in cancer chemoprotection or against other radical-mediated pathologic states.

ACKNOWLEDGMENTS

We are indebted to our coworkers and collaborators: N. Davidson, P. Dolan, P. Egner, J. Groopman, B. Roebuck, J. Swauger, P. Talalay, M. Trush, and J. Zweier. Financial support was provided by grants from the National Institutes of Health (CA 39416, CA 44530 and ES 00454) and the American Cancer Society (SIG-3) T.W.K. is recipient of Research Career Development Award CA 01230.

REFERENCES

Barch, D. H., and Fox, C. C. (1988). Selective inhibition of methylbenzylnitrosamine-induced formation of esophageal O[6]-methylguanine by dietary ellagic acid in rats. *Cancer Res.,* 48:7088–92.

Benson, A. M., Batzinger, R. P., Ou, S.-Y. L., Bueding, E., Cha, Y.-N., and Talalay, P. (1978). Elevation of hepatic glutathione *S*-transferase activities and protection against mutagenic metabolites of benzo(a)pyrene by dietary antioxidants. *Cancer Res.* 38:4486–95.

Benson, A. M., Cha, Y.-N., Bueding, E., Heine, H. S., and Talalay, P. (1979). Elevation of extrahepatic glutathione *S*-transferase and epoxide hydratase activities by 2(3)-*tert*-butyl-4-hydroxyanisole. *Cancer Res.* 39:2971–77.

Bertram, J. S., Kolonel, L. N., and Meyskens, F. L., Jr. (1987). Rationale and strategies for chemoprevention of cancer in humans. *Cancer Res.* 47:3012–31.

Bloom, A. D., Spatz, L., and Paul, N. W., eds. (1989). *Genetic Susceptibility to Environmental Mutagens and Carcinogens*, White Plains, NY: March of Dimes, pp. 1–93.

Boone, C. W., Kelloff, G. J., and Malone, W. E. (1990). Identification of candidate cancer chemopreventive agents and their evaluation in animal models and human clinical trials: a review. *Cancer Res.*, 50:2–9.

Cabral, J. R. P., and Neal, G. E. (1983). The inhibitory effects of ethoxyquin on the carcinogenic action of aflatoxin B_1 in rats. *Cancer Lett.*, 19:125–32.

Cathcart, R., Schwiers, E., Saul R., and Ames, B. (1984). Thymine glycol and thymidine glycol in human and rat urine: a possible assay for oxidative DNA damage. *Proc. Natl. Acad. Sci. USA*, 81:5633–37.

Cerutti, P. A. (1985). Prooxidant states in tumor promotion. *Science*, 227:375–81.

Cha, Y.-N, and Heine, H. (1982). Comparative effects of dietary administration of 2(3)-*tert*-butyl-4-hydroxyanisole and 3,5-di-*tert*-butyl-4-hydroxytoluene on several hapatic enzyme activities in mice and rats. *Cancer Res.* 42:2609–15.

Ch'ih, J. J., Biedrzycka, D. W., Lin, T. Khoo, M. O., and Devlin, T. (1989). 2(3)-*tert*-butyl-4-hydroxyanisole inhibits oxidative metabolism of aflatoxin B_1 in isolated rat hepatocytes. *Proc. Soc. Exp. Biol. Med.*, 192:35–42.

Chung, F.-L., Wang, M., Carmella S. G., and Hecht, S. S. (1986). Effects of butylated hydroxyanisole on the tumorigenicity and metabolism of N-nitrosodimethylamine and N-nitrosopyrrolidine in A/J mice. *Cancer Res.*, 46:165–68.

Clapp, N. K., Boweles, N. D., Satterfield, L. C., and Klima, W. C. (1979). Selective protective effect of butylated hydroxytoluene against 1.2-dimethylhydrazine carcinogenesis in BALB/c mice. *J. Natl. Cancer Inst.*, 63:1081–87.

Clapp, N. K., Tyndall, R. L. Satterfield, L.C., Klima, W. C., and Bowles, N. D. (1978). Selective sex-related modification of diethylnitrosamine-induced carcinogenesis BALB/c mice by concomitant administration of butylated hydroxytoluene. *J. Natl. Cancer Inst.*, 61:177–82.

Daoud, A. H., and Griffin, A. C. (1980). Effect of retinoic acid, butylated hydroxytoluene, selenium and ascorbic acid on azo-dye hepatocarcinogenesis. *Cancer Lett.*, 9:299–304.

Davidson, N. E., Egner, P. A., and Kensler, T. W. (1990). Transcriptional control of glutathione *S*-transferase gene expression by the chemoprotective agent

5-(2-pyrazinyl)-4-methyl-1,2-dithiole-3-thione [Oltipraz] in rat liver. *Cancer Res.*, 50:2251–55.

Dizdaroglu, M., and Bergtold, D. (1986). Characterization of free radical-induced base damage in DNA at biologically relevant levels. *Anal. Biochem.*, 156:182–88.

Fisher, S. M., Baldwin, J. K., and Adams, L. M. (1986). Effects of anti-promoters and strain of mouse on tumor promoter-induced oxidants in murine epidermal cells. *Carcinogenesis*, 7:915–18.

Greenblatt, M. (1973). Ascorbic acid blocking aminopyrine nitrosation in MZO/B_1 mice. *J. Natl. Cancer Inst.*, 50:1055–56.

Groopman, J. D., Cain, L. G., and Kensler, T. W. (1988). Aflatoxin exposure in human populations: measurements and relationships to cancer. *Crit. Rev. Toxicol.*, 19:113–45.

Guyton, K. Z., Bhan, P., Kuppusamy, P., Zweir, J. L., Trush, M. A., and Kensler, T. W. (1991). Free radical-derived quinone methide mediates skin tumor promotion by butylated hydroxytoluene: Expanded role for electrophiles in multistage carcinogenisis. *Proc. Natl. Acad. Sci. USA*, 88:946–50.

Henderson, R. F., Bechtold, W. E., Bond, J. A., and Sun, J. D. (1989). The use of biological markers in toxicology. *Crit. Rev. Toxicol.*, 20:65–82.

Inai, K., Kobuke, T., Nambu, S., Takemoto, T., Kou, E., Nishina, H. et al. (1988). Hepatocellular tumorigenicity of butylated hydroxytoluene administered orally to B6C3F_1 mice. *Jpn. J. Cancer Res.* 79:49–58.

Ito, N., Fukushima, S., and Tsuda, H. (1985). Carcinogenicity and modification of the carcinogenic response by BHA, BHT, and other antioxidants. *Crit. Rev. Toxicol.*, 15:109–50.

Ito, N., and Hirose, M. (1987). The role of antioxidants in chemical carcinogenesis. *Jpn. J. Cancer Res.*, 78:1011–26.

Kahl, R. (1984). Synthetic antioxidants: biochemical actions and interference with radiation, toxic compounds, chemical mutagens and chemical carcinogens. *Toxicology*, 33:185–228.

Kamm, J. J., Dashman, T., Conney, A. H., and Burns, J. J. (1973). Protective effects of ascorbic acid on hepatotoxicity caused by nitrate plus aminopyrine. *Proc. Natl. Acad. Sci. USA*, 70:747–49.

Kennedy, C. H., Maples, K. R., and Mason, R. P. (1990). *In vivo* detection of free radical metabolites. *Pure and Applied Chemistry*, 62:295–99.

Kensler, T. W., Egner, P. A., Dolan, P. M., Groopman, J. D., and Roebuck, B. D. (1987). Mechanism of protection against aflatoxin tumorigenicity in rats fed 5-(2-pyrazinyl)-4-methyl-1,2-dithiol-3-thione (oltipraz) and related 1,2-dithiol-3-thiones and 1,2-dithiol-3-ones. *Cancer Res.*, 47:4271–77.

Kensler, T. W., Egner, P. A., Trush, M. A., Bueding E., and Groopman, J. D. (1985). Modification of aflatoxin B_1 binding to DNA *in vivo* in rats fed phenolic antioxidants, ethoxyquin and a dithiolthione. *Carcinogenesis*, 6:759–63.

Kensler, T. W., and Taffe, B. G. (1986). Free radicals in tumor promotion. *Adv. Free Radic. Biol Med.*, 2:347–87.

Kensler, T. W., and Trush, M. A. (1985). Oxygen free radicals in chemical carcinogenesis. In L. W. Oberley, ed. *Superoxide Dismutase.* Vol III, Boca Raton: CRC Press, pp. 191–236.

Kozumbo, W. J., and Cerutti, P. (1986). Antioxidants as antitumor promoters. In D. M. Shankel, P. E. Hartman, T. Kada, and A. Hollaender, eds., *Anti-*

mutagenesis and Anticarcinogenesis Mechanisms, New York: Plenum Press, pp. 491–506.

Kozumbo, W. J., Seed, J. L., and Kensler, T. W. (1983). Inhibition by 2(3)-*tert*-butyl-4-hydroxyanisole and other antioxidants of epidermal ornithine decarboxylase activity by 12-0-tetradecanoylphorbol-13-acetate. *Cancer Res.*, 43:2555–59.

Kozumbo, W. J., Trush, M. A., and Kensler, T. W. (1985). Are free radicals involved in tumor promotion? *Chem. Biol. Interac.*, 54:199–207.

Malkinson, A. M., and Beer, D. S. (1984). Pharmacologic and genetic studies on the modulatory effects of butylated hydroxytoluene on mouse lung adenoma formation. *J. Natl. Cancer Inst.*, 73:925–33.

Mandal, S., Ahuja, A., Shivapurkar, M., Cheng, S. J., Groopman, J. D., and Stoner, G. D. (1987). Inhibition of aflatoxin B_1 mutagenesis in *Salmonella typhimurium* and DNA damage in cultured rat and human tracheobronchial tissues by ellagic acid. *Carcinogenesis*, 8:1651–56.

Mandal, S., Shivapurkar, M., Galati, A. J., and Stoner, G. D. (1988). Inhibition of N-nitrosobenzylmethylamine metabolism and DNA binding in cultured rat esophagus by ellagic acid. *Carcinogenesis*, 9:1313–16.

Mandel, H. G., and Cohn, V. H. (1985). Fat-soluble vitamins. In A. G. Gilman, L. S. Goodman, T. W. Rall, and F. Murad, eds. *The Phamacological Basis of Therapeutics*. New York: MacMillan, pp. 1573–91.

Mandel, H. G., Mansom, M. M., Judah, D. J., Simpson, J. L., Green, J. A., Forrester, L. M. et al. (1987). Metabolic basis for the protective effect of the antioxidant ethoxyquin on aflatoxin B_1 hepatocarcinogenesis in the rat. *Cancer Res.*, 47:2518–23.

Marnett, L. J., (1987). Peroxyl free radicals: potential mediators of tumor initiation and promotion. *Carcinogenesis*, 8:1365–73.

McCormick, D. L., Major, N., and Moon, R. C. (1984). Inhibition of 7,12-dimethylbenz(a)anthracene-induced rat mammary carcinogenesis by concomitant or postcarcinogen antioxidant exposure. *Cancer Res.*, 44:2858–63.

Miller, J. A. (1970). Carcinogenesis by chemicals: an overview—G.H.A. Clowes memorial lecture. *Cancer Res.*, 30:559–76.

Mirvish, S. S., Wallcave, L., Eagen, M., and Shubik, P. (1972). Ascorbate-nitrate reaction: possible means of blocking the formation of carcinogenic N-nitroso compounds. *Science*, 177:65.

Mizumoto, K., Ito, S., Kitazawa, S., Tsutsumi, M., Denda, A., and Konishi, Y. (1989). Inhibitory effect of butylated hydroxyanisole administration on pancreatic carcinogenesis in Syrian hamsters initiated with N-nitrosobis(2-oxopropyl)amine. *Carcinogenesis*, 10:1491–94.

O'Connell, J. F., Klein-Szanto, A. J. P., DiGiovanni, D. M., Freis, J. W., and Slaga, T. J. (1986). Enhanced malignant progression of mouse skin tumors by the free-radical generator benzoyl peroxide. *Cancer Res.*, 46:2863–65.

Perchellet, J. P., Owen, M. D., Posey, T. D., Orten, D. K., and Schneider, B. A. (1985). Inhibitory effects of glutathione level-raising agents and D-α-tocopherol on ornithine decarboxylase induction and mouse skin tumor promotion by 12-0-tetradecanoylphorbol-13-acetate. *Carcinogenesis*, 6:567–73.

Perchellet, J. P., and Perchellet, E. M. (1989). Antioxidants and multistage carcinogenesis in mouse skin. *Free Radic. Biol. Med.* 7:377–408.

Preuss-Schwartz, D., Nimesheim, A., and Marnett, L. J. (1989). Peroxyl radical- and cytochrome P-450-dependent metabolic activation (+)-7,8-dihydroxy-

7,8-dihydrobenzo(a)pyrene in mouse skin *in vitro* and *in vivo*. *Cancer Res.*, 49:1732–37.

Prochaska, H. J., Bregman, H. S., DeLong, M. J., and Talalay, P. (1985). Specificity of induction of cancer protective enzymes by analogues of *tert*-butyl-4-hydroxyanisole (BHA). *Biochem. Pharmacol.*, 34:3909–14.

Prochaska, H. J., and Talalay, P. (1988). Regulatory mechanisms of monofunctional and bifunctional anticarcinogenic enzyme inducers. *Cancer Res.*, 48:4776–82.

Reddy, B. S., and Maeura, Y. (1984). Dose-response studies of the effect of dietary butylated hydroxyanisole on colon carcinogenesis induced by methylazoxymethanol acetate in female CF1 mice. *J. Natl. Cancer Inst.*, 72:1181–87.

Roebuck, B. D., Liu, Y.-L., Rogers, A. E., Groopman, J. D., and Kensler, T. W. (1991). Protection against aflatoxin B_1-induced hepatocarcinogenesis in F344 rats by 5-(2-pyrazinyl)-4-methyl-1, 2-dithiole-3-thione (Oltipraz): predictive role for short-term molecular dosimetry. *Cancer Res.*, 51:5501–06.

Rotstein, J. B., and Slaga, T. J. (1988). Effect of exogenous glutathione on tumor progression in the murine skin multistage carcinogenesis model. *Carcinogenesis*, 9:1547–51.

Ruch, R. J., and Klaunig, J. E. (1986). Antioxidant prevention of tumor promoter induced inhibition of mouse hepatocyte intracellular communication. *Cancer Lett.*, 33:137–50.

Salocks, C. B., Hsieh, D. P. H., and Byard, J. L. (1984). Effects of butylated hydroxytoluene pretreatment on the metabolism and genotoxicity of aflatoxin B_1 in primary cultures of adult rat hepatocytes: selective reduction of nucleic acid binding. *Toxicol. Appl. Pharmacol.*, 76:498–509.

Seed, J. L., Nakamura, Y., and Colburn, N. H. (1987). Implication of superoxide radical anion in promotion of neoplatic transformation in mouse JB6 cells by TPA. In P. A. Cerutti, O. F. Nygaard, and M. G. Simic, eds. *Anticarcinogenesis and Radiation Protection*. New York: Plenum Press, pp. 175–81.

Siedegård, J., Pero, R. W., Miller, D. G., and Beattie, E. J. (1986). A glutathione transferase in human leukocytes as a marker for the susceptibility to lung cancer. *Carcinogenesis*, 7:751–53.

Shigenaga, M., Gimeno, C. J., and Ames, B. N. (1989). Urinary 8-hydroxy-2′-deoxyguanosine as a biological marker of *in vivo* oxidative DNA damage. *Proc. Natl. Acad. Sci. USA*, 86:9697–701.

Slaga, T. J., and Bracken, W. M. (1977). The effects of antioxidants on skin tumor initiation and aryl hydrocarbon hydroxylase. *Cancer Res.*, 37:1631–35.

Slaga, T. J., Klein-Szanto, A. J. P., Triplett, L. L., and Yotti, L. P. (1981). Skin tumor-promoting activity of benzoyl peroxide, a widely used free radical-generating compound. *Science*, 213:1023–25.

Smart, R. C., Huang, M.-T., Han, Z. T., Kaplan, M. C., Focella, A., and Conney, A. H. (1987). Inhibition of 12-0-tetradeconylphorbol-13-acetate induction of ornithine decarboxylase activity, DNA synthesis, and tumor promotion in mouse skin by ascorbic acid and ascorbyl palmitate. *Cancer Res.*, 47:6633–38.

Swauger, J. E., Dolan, P. M., and Kensler, T. W. (1990). Role of free radicals in tumor promotion and progression by benzoyl peroxide. In M. L. Mendelsohn and R. J. Albertini, eds. *Mutation and the Environment. Part D: Carcinogenesis*. New York: Wiley-Liss, pp. 143–52.

Taffe, B. G., and Kensler, T. W. (1988). Tumor promotion by a hydroperoxide metabolite of butylated hydroxytoluene, 2,6-di-*tert*-butyl-4-hydroperoxyl-4-methyl-2,5-cyclohexadienone, in mouse skin. *Res. Commun. Chem. Pathol. Pharmacol.*, 61:291–303.

Taffe, B. G., Zweier, J. L., Pannell, L. K., and Kensler, T. W. (1989). Generation of reactive intermediates from the tumor promoter butylated hydroxytoluene hydroperoxide in isolated murine keratinocytes or by hematin. *Carcinogenesis*, 10:1261–68.

Talalay, P., DeLong, M. J., and Prochaska, H. J. (1987). Molecular mechanisms in protection against carcinogenesis. In J. G. Cory and A. Szentivani, eds. *Cancer Biology and Therapeutics*. New York: Plenum Press, pp. 197–216.

Talalay, P., DeLong, M. J., and Prochaska, H. J. (1988). Identification of a common chemical signal regulating the induction of enzymes that protect against chemical carcinogenesis. *Proc. Natl. Acad. Sci. USA*, 85:8261–65.

Thompson, J. A., Schullek, K. M., Fernandez, C. A., and Malkinson, A. M. (1989). A metabolite of butylated hydroxytoluene with potent tumor promoting activity in mouse lung. *Carcinogenesis*, 10:773–75.

Trush, M. A., Seed, J. L., and Kensler, T. W. (1985). Reactive oxygen-dependent metabolic activation of polycyclic aromatic hydrocarbons by phorbol ester-stimulated human polymorphonuclear leukocytes: possible link between inflammation and cancer. *Proc. Natl. Acad. Sci. USA*, 82:5194–98.

Ulland, B. M., Weisburger, J. H., Yamamoto, R. S., and Weisburger, E. K. (1973). Antioxidants and carcinogenesis: butylated hydroxytoluene, but not diphenyl-p-phenylenediamine inhibits cancer induction by N-2-fluorenylacetamide and by N-hydroxy-N-2-fluorenylacetamide in rats. *Food Chem. Toxicol.*, 11:199–207.

Wattenberg, L. W. (1972a). Inhibition of carcinogenic effects of diethylnitrosamine and 4-nitroquinoline-N-oxide by antioxidants. *Fed. Proc.*, 31:633.

Wattenberg, L. W. (1972b). Inhibition of carcinogenic and toxic effects of polycyclic hydrocarbons by phenolic antioxidants and ethoxyquin. *J. Natl. Cancer Inst.*, 48:1425–30.

Wattenberg, L. W. (1973). Inhibition of chemical carcinogen-induced pulmonary neoplasia by butylated hydroxyanisole. *J. Natl. Cancer Inst.*, 50:1541–44.

Wattenberg, L. W., (1980). Inhibition of chemical carcinogenesis by antioxidants. In T. J. Slaga, ed. *Carcinogenesis, Vol. 5: Modifiers of Chemical Carcinogenesis*. New York: Raven, pp. 85–98.

Wattenberg, L. W. (1985). Chemoprevention of cancer. *Cancer Res.*, 45:1–8.

Wattenberg, L. W., and Bueding, E. (1986). Inhibitory effects of 5-(2-pyranzinyl)-4-methyl-1,2-dithiol-3-thione (Oltipraz) on carcinogenesis induced by benzo(a)pyrene, diethylnitrosamine and uracil mustard. *Carcinogenesis*, 7:1379–81.

Wattenberg, L. W. and Lam, L. K. T. (1980). Inhibition of chemical carcinogenesis by phenols, coumarins, aromatic isothiocyanates, flavones, and indoles. In M. S. Zedeck, and M. Lipkin, eds. *Inhibition of Tumor Induction and Development*. New York: Plenum Press, pp. 1–22.

Wattenberg, L. W., Hochalter, J. B., Prabhu, U. D. G., and Galbraith, A. R. (1987). Nucleophiles as anticarcinogens. In P. A. Cerutti, O. F. Nygaard, and M. G. Simic, eds. *Anticarcinogenesis and Radiation Protection*. New York: Plenum Press, pp. 233–40.

Weitberg, A. B., Weitzman, S. A., Clark, E. P., and Stossel, T. P. (1985). Effects of antioxidants on oxidant-induced sister chromatid exchange formation. *J. Clin. Invest.*, 75:1835–41.

Williams, G. M., Maeura, Y., and Weisburger, J. H. (1983). Simultaneous inhibition of liver carcinogenicity and enhancement of bladder carcinogenicity of N-2-fluorenylacetamide by butylated hydroxytoluene. *Cancer Lett.*, 19:55–60.

Williams, G. M., Tanaka, T., and Maeura, Y. (1986). Dose-related inhibition of aflatoxin B_1 induced hepatocarcinogenesis by the phenolic antioxidants, butylated hydroxyanisole and butylated hydroxytoluene. *Carcinogenesis*, 7:1043–50.

Witschi, H. P., and Lock, S. (1979). Enhancement of adenoma formation in mouse lung by butylated hydroxytoluene. *Toxicol. Appl. Pharmacol.*, 50:391–400.

Wogan, G. N., and Gorelick, N. J. (1985). Chemical and biochemical dosimetry of exposure to genotoxic chemicals. *Environ. Health Perspect.*, 62:5–18.

Wood, A. W., Huang, M. T., Chang, R. L., Newmark, H. L., Lehr, R. E., Yagi, H. et al. (1982). Inhibition of the mutagenicity of bay-region diolepoxides of polycyclic aromatic hydrocarbons by naturally occurring plant phenols: exceptional activity of ellagic acid. *Proc. Natl. Acad. Sci. USA*, 79:5513–17.

Wurtzen, G., and Olsen, P. (1986). Chronic study on BHT in rats. *Food Chem. Toxicol.*, 24:1121–25.

Yang, C. S., Sydor, W., Jr., Martin, M. B., and Lewis, K. F. (1981). Effects of butylated hydroxyanisole on the aryl hydrocarbon hydroxylase of rats and mice. *Chemic. Biol. Interact.*, 37:337–50.

Zedeck, M. S., Sternberg, S. S., McGowan, J., and Poynter, R. W. (1972). Methylazoxymethanol acetate: induction of tumors and early effects of RNA synthesis. *Fed. Proc.*, 31:1485–92.

Index

Page numbers in italics indicate figures or tables.